Severe and Hazardous Weather in Canada
The Geography of Extreme Events

Severe and Hazardous Weather in Canada
The Geography of Extreme Events

Cathy T. Conrad

OXFORD
UNIVERSITY PRESS

8 Sampson Mews, Suite 204, Don Mills, Ontario M3C 0H5
www.oupcanada.com

Oxford University Press is a department of the University of Oxford.
It furthers the University's objective of excellence in research, scholarship,
and education by publishing worldwide in

Oxford   New York

Auckland   Cape Town   Dar es Salaam   Hong Kong   Karachi
Kuala Lumpur   Madrid   Melbourne   Mexico City   Nairobi
New Delhi   Shanghai   Taipei   Toronto

With offices in

Argentina   Austria   Brazil   Chile   Czech Republic   France   Greece
Guatemala   Hungary   Italy   Japan   Poland   Portugal   Singapore
South Korea   Switzerland   Thailand   Turkey   Ukraine   Vietnam

Oxford is a trade mark of Oxford University Press
in the UK and in certain other countries

Published in Canada
by Oxford University Press

Copyright © Oxford University Press Canada 2009

The moral rights of the author have been asserted

Database right Oxford University Press (maker)

First published 2009

All rights reserved. No part of this publication may be reproduced,
stored in a retrieval system, or transmitted, in any form or by any means,
without the prior permission in writing of Oxford University Press,
or as expressly permitted by law, or under terms agreed with the appropriate
reprographics rights organization. Enquiries concerning reproduction
outside the scope of the above should be sent to the Rights Department,
Oxford University Press, at the address above.

You must not circulate this book in any other binding or cover
and you must impose this same condition on any acquirer.

Library and Archives Canada Cataloguing in Publication Data

Conrad, Catherine Treena, 1971–
Severe and hazardous weather in Canada : the geography of extreme events / Catherine Conrad

Includes index.   ISBN 978-0-19-542627-4

1. Severe storms—Canada.   2. Climatic extremes—Canada.   3. Climatic changes—Canada.
4. Canada—Climate.   I. Title

QC985.C67   2008   551.6971   C2008-903808-8

This book is printed on permanent (acid-free) paper ∞.
Printed and bound in Canada

Front Cover Image: Michael Orton/Getty Images
Back Cover Image: Soubrette/iStockphoto

Contents

Foreword   ix

Acknowledgements   xi

List of Figures   xiii

List of Tables   xvii

Chapter 1 Introduction to Severe and Hazardous Weather in Canada   1
    Introduction   1
    The Context of Natural Hazards   3
        Natural *Disasters* Versus Natural *Hazards*   3
    How is Severe Weather Defined and Classified?   5
    Introduction to Atmospheric Hazards in Canada   9
    Canadian Disaster Mitigation, Preparedness, Response,
        and Recovery Activities   13
    Link to Climate Change and Natural Climatic Oscillations   15
    Chapter Summary   19

Chapter 2 Severe Weather Forecasting   21
    Introduction   21
    Weather Forecast Quality   22
    The Way It Was   23
    The Forecasting Procedure   24
        Data Acquisition   25
        Weather Forecasting Models   26
        The Canadian 'Ensemble Prediction System'   28
    Radar Technology   28
    Satellite Imagery   31
    The Canadian Lightning Detection Network   33
    Seasonal Forecasts   34
    How and What Warnings Are Issued   35
    'Citizen Forecasts'   38
    Forecasting to the North   39
    Chapter Summary   40

Chapter 3 Severe 'Winter' Weather   41
    Introduction   41
    Causes of Storms   43
    Winter Storm Tracks   45
    Wreckhouse Winds   48
    Alberta Clippers   49
    Nor'easters   50
    Chinook Winds   51
    Snow   53
        Notable Canadian Snowstorms   56
    Blizzards   58
        Notable Canadian Blizzards   61
    Ice Storms   62
        The 1998 Ice Storm   65
    Cold and Wind Chill   67
        So Cold That Niagara Falls Freezes?   71
        Visualizing the Cold   72
    Black Ice   72
    Winter Weather and a Changing Climate   73
    Looking Ahead   74
    Chapter Summary   76

Chapter 4 Severe 'Summer' Weather   77
    Introduction   77
    Summer Storm Tracks   78
    Heat   79
        Urban Heat Island Effect   82
    Humidity   83
    Smog   85
    Thunderstorms   88
    Hail   89
    Tornadoes   92
    Lightning   100
    Climate Change and Severe Summer Weather   101
    Chapter Summary   104

Chapter 5 Water: Too Much or Not Enough   105
    Introduction   105
    Heavy Precipitation Events   106
        Processes   108
    Floods   109
        Causes and Examples of Flooding   113
        Flood Management   116
    Fog   120
        Processes and Types of Fog   123
        Impacts of Fog in Canada   125

Drought   127
   Severe Canadian Droughts   129
   Monitoring and Drought Indices   130
Climate Change and Precipitation in Canada   131
Climate Change and Fog   133
Climate Change and Implications for Drought   133
Chapter Summary   135

Chapter 6 Tropical Cyclones   137
Introduction   137
Processes   139
Classification   141
Naming Hurricanes   143
Impacts of Tropical Cyclones   144
   Wind   144
   Rainfall   145
   Storm Surge   146
   Tornadoes   147
The Canadian Hurricane Centre: Forecasting Tropical Storms   147
Extended-Range Forecasts   151
Tropical Storms Affecting Canada   152
Climate Change and Tropical Cyclones   166
Chapter Summary   168

Chapter 7 Canadians and Weather: Vulnerability, Risk, Adaptation, and Mitigation   169
Introduction   169
Vulnerability   170
Resilience   172
Perception of Risk   173
Mitigation and Adaption   176
The Context for Canada   179
Conclusion   181

References   183

Index   200

Foreword

On 29 September 2003, Hurricane Juan reawakened Nova Scotians and Prince Edward Islanders to the importance of preparedness for severe weather. The population centres of Halifax and Charlottetown sustained some of Juan's worst impacts, and for them, it was a true 'storm of the century.' Yet despite clear, consistent warnings, few people were truly prepared for that storm. When the dust had settled—and 100 million trees had fallen—we found that the hurricane had served up another reminder that severe weather preparedness must include more than accurate and timely weather forecasts; it must also include awareness and knowledge of the threatening hazard. Accordingly, books that raise our awareness of the realistic threats of weather hazards go a long way towards mitigating their impact. This is such a book.

From its breadth, covering the full plethora of Canadian severe weather, to its depth, highlighting many of the current and contentious debates about weather and climate, this text offers a full-spectrum introduction to severe weather in the Great White North. As stated by the author, Catherine Conrad, the purpose of the book is to unravel the patterns of severe weather in Canada to help us understand the type, magnitude, location, and timing of the various natural weather hazards. She appeals to the reader by drawing attention to the various lessons taught by our seemingly countless weather disasters. Despite being targeted as a university text, this book can easily be seen as appealing both to students and to interested readers beyond the academic community.

The chapter layouts are effective: objectives—introduction—details—summary. In other words, the author first tells you what she's going to tell you . . . then she tells you why she's going to tell it to you . . . then she tells it to you . . . finally she tells you what she just told you. This is a helpful roadmap for students journeying through the text. The strength of the book lies in its unapologetic foundation: that Canadians suffer more weather disasters than necessary. As meteorologists teach to all who will listen, weather hazards do not have to become weather disasters; the quiet—but deadly— link of vulnerability lies well within our own hands. Remove vulnerability and a hazard will never become a disaster. As the author muses, the great ice storm of 1998 would have been a mild curiosity 150 years earlier, yet the technology and infrastructure around which we have built our lives put us in harm's way in 1998; we created our own vulnerability. The discipline of natural hazard risk incorporates the notions of both hazard and vulnerability. We each need to assess our own vulnerability to hazardous weather, and this text lays the foundation to educate the reader on matters that are common to all Canadians because every one of us is affected by the hazard of severe weather.

The reader will appreciate the manner in which the information is presented with balanced emotion. This comes out well in the peripheral topics of weather forecasting and climate change—topics that can easily evoke passionate homilies from armchair meteorologists. The author's treatment of these subjects in particular reveals a respect for their complexities and an honest assessment of our current understanding. I must confess to wondering how a non-meteorologist could possibly understand what we do, how we do it, and the myriad of challenges that assault forecasters on an hourly basis. Somehow, without being flippant about our skill or dismissive of our challenges, she portrays an accurate state of our *art* and *science*. This is surprisingly refreshing to find in a text written by someone who is not 'one of us.'

Whether *lessons taught* by our hazardous weather will transform into *lessons learned* remains to be seen. Perhaps students studying from this text will become tomorrow's important decision-makers who will ensure that the lessons lead us to making the right changes through a mindset of mitigation rather than through the pain and shame of recovery. Reducing our vulnerability is clearly an important task, but in the spirit of award-winning author Stephen Covey, we need to do so before it becomes urgent. Everyone recognizes that decisions taken under an urgent paradigm tend to be less effective and almost always short-sighted. The author is careful not to preach against our propensity for building vulnerability into our lives and infrastructure. Rather, she allows the reader to come to the obvious conclusion through an avalanche of data, anecdotes, flashbacks, and historical perspectives.

Peter J. Bowyer
Program Supervisor, Canadian Hurricane Centre,
Environment Canada

The opinions expressed in the above foreword are solely those of Mr Bowyer. They do not constitute an endorsement of the contents and opinions of the author of the book by either Environment Canada or the Government of Canada

Acknowledgements

I owe a huge debt of thanks to a number of people who helped this book come to fruition and be as thorough and comprehensive as possible. Thank you to Jennifer Charlton, Katherine Skene, and the editorial team at Oxford University Press for your assistance at every stage in the process of the text, from pitching the idea through to the final publication. You are a very hard-working and dedicated group of individuals. Thank you to Judith Turnbull, who edited the text and whose hard work and diligence are second to none. Thank you to Don Bonner, the cartographer in the Department of Geography at Saint Mary's University, for preparing the maps throughout the text. I also need to thank Peter Bowyer, program supervisor at the Canadian Hurricane Centre, for his incredible assistance in the form of providing comments and reviews on all of the chapters, but for his assistance on the tropical cyclone section in particular. Other individuals assisted in small but important ways, including Tim Bullocks and Bill Richards from Environment Canada and Gary Lines for reviewing sections on climate change, Dan Shrubsole for his paper on flood management in Canada, James Morrison for papers on nineteenth-century weather forecasting, and Alan Ruffman for papers on historical tropical cyclones. I thank the three anonymous reviewers whose input and feedback on the first draft of the text were much appreciated and enabled the final version to be greatly improved. I would also like to thank my students in the Weather and Climate course (GEOG 3343) in the winter semester of 2008 for allowing me to preview sections of this text on them. Thank you to Ben Lemieux, Sarah Weston, Yukari Hori, and Amber Silver for a great job on research assignments, which greatly assisted in those sections of the text. Last and of course not least, thank you to Scott for listening to many weather-related anecdotes and to Jakob and Sam for being such great kids (which always makes one's job so much easier to accomplish!)

List of Figures

Figure 1.1 Spatial and temporal scales of weather and weather-related natural hazards.

Figure 1.2 Spatial distribution of the Climate Severity Index.

Figure 1.3 Economic damages by disaster type (US$ x 1000).

Figure 1.4 Weather-related natural disasters by province (major multiple-payment occurrences), 1987–2002.

Figure 1.5 Frequency distribution of weather-related disasters in Canada by province, 1900–2000.
Number of disasters = 473.

Figure 2.1 Location of Meteorological Service of Canada weather stations.

Figure 2.2 Location of Doppler weather radar stations covering Canada.

Figure 2.3: Marion Bridge Radar Station.

Figure 2.4: Electromagnetic Spectrum.

Figure 2.5 Location of Canadian Lightning Detection Network (CLDN) sensors.

Figure 2.6 The difference between a seasonal forecast prediction (image on left) and the observed conditions.

Figure 3.1 Spring Garden Road, Halifax: First winter storm of the season, 4 December 2006.

Figure 3.2 Types of winter storms and their most common locations in Canada.

Figure 3.3 Winter storm tracks.

Figure 3.4 Winter storm tracks that impact Nunavut and the Arctic region.

Figure 3.5 The Wreckhouse wind region of Newfoundland.

Figure 3.6 The average trajectory of an Alberta clipper.

Figure 3.7 Region of most intense chinook winds.

Figure 3.8 The aftermath of a large snowstorm in Newfoundland. Snow accumulates in deep drifts near large objects like this house.

Figure 3.9 Average annual snowfall (cm).

Figure 3.10 Average annual number of days with blowing snow in Canada.

Figure 3.11 Blizzard of 1947.

Figure 3.12 Median annual hours of freezing rain and freezing drizzle (combined), 1976–1990.

Figure 3.13 Freezing rain accumulations during the 1998 ice storm.

Figure 3.14 Niagara Falls freezes.

Figure 3.15 Winter national temperature departures and long-term trend, 1948–2007.

Figure 3.16 Canadian Climate Change Scenarios Network Bioclimate Profiles for Alert, Prince George, and St John's.

Figure 4.1 Summer storm tracks across Canada.

Figure 4.2 Smog in Toronto.

Figure 4.3 Average number of days (1987–1992) when one-hour average concentrations of ozone surpassed 82 parts per billion (ppb). An excess of 65 ppb is indicative of relatively poor air-quality conditions.

Figure 4.4 Regions of Quebec that are covered by the Info-Smog program.

Figure 4.5 Average annual number of days with hail in Canada.

Figure 4.6 Photo of hail damage.

Figure 4.7 Photo of tornado impact/damage at Elie.

Figure 4.8 Distribution and number of tornadoes across Canada per year.

Figure 5.1 General trends in precipitation across North America.

Figure 5.2 Precipitation pattern in British Columbia, with the distribution of isohyets exhibiting the results of the orographic effect as the westerly winds off the Pacific are forced across the north–south-trending Coastal-Cascade Mountains.

Figure 5.3 Source regions for North American air masses.

Figure 5.4 The distribution of floods by province and territory from 1900 to 1997.

Figure 5.5 Flood disasters in Canada, 1900–1971 (single and multiple events).

Figure 5.6 Image from Badger: town encased in ice.

Figure 5.7 Red River Floodway.

Figure 5.8 Global patterns of fog.

Figure 5.9 Average number of days per year when fog reduces visibility to less than 1 km, based on data from 1971 to1999 (Canadian overview).

Figure 5.10 Fog-prone areas on Canada's east and west coasts.

Figure 6.1 Storms of tropical origin with hurricane-force winds (1851–2003).

Figure 6.2 Canadian Hurricane Centre area of forecast responsibility and Response Zone.

Figure 6.3 Hurricane Juan approaching Nova Scotia (2003).

Figure 6.4 Preferred tropical storm tracks in the Northwest Atlantic, 1949–1983.

Figure 6.5 Tropical storm frequency by year, 1901–2000.

Figure 6.6 Atlantic tropical storm frequencies, 1901–2000.

Figure 6.7 Comparison of tropical storm regions (within black line), 18°C average, isotherm and 500 m water depth, 1949–1983.

Figure 6.8 The path of Hurricane Juan, 2003.

Figure 6.9 Destruction caused by Hurricane Juan.

List of Tables

Table 1.1 Factors that make a community more or less vulnerable to a natural hazard.

Table 1.2 Climate Severity Index (CSI) categories and subcategories.

Table 1.3 Recent Canadian weather-related natural disasters.

Table 1.4 Number of weather-related disasters and associated deaths and injuries in Canada, 1900–1999.

Table 1.5 Canada's most expensive natural disasters.
*Billions of 1999 dollars; the range reflects the differing figures from various sources.

Table 1.6 Likelihood of future climatological trends.

Table 2.1 Examples of weather warnings.

Table 2.2 What weather watchers report.

Table 3.1 Notable snowstorms in recent Canadian history.

Table 3.2 Notable blizzards in recent Canadian history.

Table 3.3 Significant freezing rain events in recent Canadian history.

Table 3.4 Some notable Canadian cold snaps in recent history.

Table 4.1 Significant heat waves in Canada, 1912–2006.

Table 4.2 The Fujita Scale.

Table 4.3 Some of the worst tornado and hail events that have struck across Canada from 1879 to 2004.

Table 5.1 Some of the heaviest rain events recorded for a single day in Canada.

Table 5.2 Canadian cities that receive more than 125 rainy days in a year.

Table 5.3 Significant floods from the nineteenth and twentieth centuries.

Table 5.4 The qualitative explanations for the Palmer Drought Severity Index values.

Table 6.1 Frequency of impacts from tropical cyclones by province (1901–2000).

Table 6.2 Some of the worst tropical cyclones in Canadian history.

Table 6.3 Storms that entered the Canadian Response Zone.

Table 6.4 Notable Canadian hurricanes.

Chapter One

Introduction to Severe and Hazardous Weather in Canada

Objectives

- To differentiate between *natural hazards* and *natural disasters*
- To introduce the concept of social, economic, and environmental vulnerability
- To explain how severe weather is classified and defined
- To introduce the variety and scope of atmospheric hazards in Canada
- To introduce Canadian disaster mitigation, preparedness, response, and recovery activities
- To explore the link between severe weather and climate change and natural climatic oscillations

Introduction

'Of all the natural hazards which threaten human society, those caused or facilitated by weather extremes are the most common. On a world-wide basis, relatively few people are directly exposed to geologic hazards . . . everyone, however, is exposed to the variability of weather and climate. Canada, like many countries, is exposed to a wide variety of weather extremes.'

Brun et al. 1997

Canada's climate is typified by extremes, and thus Canadians are interested in the weather out of necessity and concern. With the inevitable changes in our global climate, scientists as well as the general public are concerned with the impact such change will have on extreme weather events in Canada. A poll of Canadians taken in 2007 indicated that, among green issues, global warming was the greatest concern, evidently put at that level because respondents were 'freaked out' by unusual weather that they took as 'clear proof of global warming' (*IQP* 2007). Because extreme weather generally has a severe and direct impact on our everyday lives, community, and environment, detection of changes in extremes has become important in current climatological research (Vincent and Mekis 2006).

Loss of life and economic damage from severe weather and climate events have been steadily increasing since the beginning of the twentieth century in Canada (Kunkel et al. 1999b; Meehl et al. 2000a; Environment Canada 2003d). There is also considerable evidence that shifts in the frequency of extremes will occur with changes in climate, thus exposing additional segments of the population and infrastructure to increased risk (Easterling et al. 2000). Over the past decade, Canada has experienced many of its most severe natural disasters, and experts believe that even bigger and more devastating ones are inevitable. While geophysical disasters, such as earthquakes, have remained relatively constant in this country over the past 50 years, weather-related disasters have skyrocketed. Climate change is projected to exacerbate this situation in the future, as it is expected to increase the frequency and severity of some extreme weather events (Environment Canada 2003d). Media attention has increasingly focused on severe weather and on the often large loss of human life and increasing costs associated with it, thus elevating interest and concern among the general population and in academia (Easterling et al. 2000; Karl and Easterling 1999). The number of extraordinarily severe floods, storms, and other weather calamities that have occurred since the early 1990s would seem to suggest that such events are becoming more common (Francis and Hengeveld 1998), although the connection between severe weather and global warming is still open for debate. The likelihood that weather-related disasters are on the rise is supported by an analysis conducted by the Geneva Secretariat for the International Decade for Natural Disaster Reduction. Between the mid-1960s and the early 1990s, the number of all disasters increased, but the weather-related disasters increased at a much higher rate. It is difficult to provide direct evidence of a trend, given the nature of weather and climate data and the inherent natural variability made evident by historical records. The lack of long-term climate data suitable for analysis of extremes is the single biggest obstacle to quantifying whether extreme events have changed over the twentieth century, either worldwide or on a more regional basis (Easterling et al. 1999, 2000).

Severe weather can be blamed for spells of 'cabin fever' or general monotony, not to mention the events that can accompany severe weather, such as power outages, motor vehicle accidents, and, in the most extreme cases, even the destruction of homes and societal infrastructure. There is little surprise, then, that Canadians are weather obsessed. Perhaps some might even be described as having severe weather phobia, defined as 'an intense, debilitating, unreasonable fear of severe weather' (Westeveld 1996; Westeveld et al. 2006). We are a nation of weather extremes, and with the vastness of the nation, it isn't difficult to understand why. But why do particular severe weather events occur where they do? What are some of the most extreme events that have happened in the past and what might Canadians expect from severe weather in the future? What do patterns of severe weather look like across the country, and what accounts for those spatial trends? We are all affected by the weather; it can influence us in an everyday way, such as deciding whether to carry an umbrella or not, all the way to life-and-death decisions, such as whether to evacuate our homes in a time of disaster.

Weather is a topic that can occupy national headlines. In many ways, it defines our nation. This book uncovers the patterns of severe weather in Canada, both in time and space, and endeavours to help us understand events that affect our lives and livelihoods. The type and magnitude of natural disasters and community vulnerabilities vary considerably across Canada (Etkin et al. 2004). This book attempts to unravel those patterns.

The Context of Natural Hazards

'While many people are aware of the terrible impact of disasters throughout the world, few realize that this is a problem that we can do something about.'

Kofi Annan, former Secretary-General, United Nations

Natural *Disasters* Versus Natural *Hazards*

'Natural disasters are the extreme of natural hazards and occur when social vulnerability is triggered by an extreme event.'

Street et al. 1997

Researchers differentiate between the terms *natural disaster* and *natural hazard*. The term *disaster* is often reserved for those cases where humans and their infrastructure are impacted. In many cases, the extreme events that cause human suffering are rejuvenating for the natural environment. Forest fires replenish soil fertility, and hurricanes bring nutrients to coastal wetlands. However, natural phenomena like hurricanes, floods, earthquakes, and tornadoes can be hazards that have the potential to harm people and damage property. These hazards only become disasters when they interact with vulnerable communities in a way that overwhelms the communities' ability to cope. Natural disasters occur when a hazard triggers vulnerability and the damage is so extensive that the affected community cannot recover through the use of its own resources (Cannon 1994). *Vulnerability* refers to the likelihood that a community will suffer injuries, deaths, or property damage from a hazardous event. It is a measure of how well prepared and equipped the community is to avoid or cope with such events (Etkin et al. 2004). There is little doubt that society as a whole has become more vulnerable to extreme weather (Kunkel et al. 1999b). Population and infrastructure continue to increase in areas that are vulnerable to such extremes as flooding, storm damage, and severe heat or cold (Easterling et al. 2000). The vulnerability and what disaster analysts call the 'risk burdens' of communities and countries are increasing as a result of a variety of everyday development decisions. Populations are too often concentrated in risky areas such as floodplains, for example, and the destruction of forests and wetlands is harming the capacity of the environment to withstand hazards. Looming above all this is the threat of global climate change and rising sea levels, the result of increased greenhouse gas concentrations in the atmosphere caused by human activity (United Nations 2004). The discipline of natural hazard risk incorporates the notions of both hazard and vulnerability.

Over the past century, we have come to understand that disasters are the disruptive and/or deadly and destructive outcome of triggering agents that have interacted with, and are exacerbated by, various forms of vulnerability (McEntire 2001). Natural disasters were believed to be 'acts of God' in centuries past. It is now understood that human decisions play a significant role in determining our vulnerability and capacity to cope with the consequences of extreme events. From this perspective, two interacting variables—the *hazard* (triggering agent) that threatens a community and the *vulnerability* of the affected population—can determine whether or not an emergency becomes a disaster (McBean and Henstra 2003).

Natural hazards can affect anyone, anywhere. 'People are threatened by hazards because of their social, economic and environmental vulnerability, which must be taken into account if sustainable development is to be achieved. Disaster risk reduction therefore concerns everyone, from villagers to heads of state, from bankers and lawyers to farmers and foresters, from meteorologists to media chiefs' (Annan 2004, viii). We have been reminded of just how vulnerable we are in recent years, both in developed and developing nations. Record high temperatures were recorded all across western Europe in April 2007, with the British Meteorological Office declaring that the 12 months ending in April 2007 were the warmest in nearly 350 years, when records of temperature (known as the Central England Temperature) began to be kept. In 2005, Europe suffered the worst floods it had experienced for centuries, while Australia and parts of Brazil, Portugal, and Spain were hit by serious droughts. Mumbai, India, experienced 944 mm of precipitation within 24 hours on 26 July 2005, resulting in the most expensive insurance loss in India to date. Tropical cyclones hit Mauritius, Réunion, the Republic of Korea, Japan, and Mexico, and tornadoes and hurricanes left a trail of devastation in the United States. The insurance giant Munich Re counted 650 natural catastrophes in 2005, calculated losses of life as in excess of 100,000 people, and estimated the economic losses at a record-breaking US$210 billion. Twenty-eight Atlantic tropical storms and hurricanes developed in 2005, breaking all meteorological and monetary records. In 2006, however, the Atlantic hurricane season failed to be as active as forecasters had predicted, with only 10 named storms and US$500 million in losses. In 2006–7, eight hydro-meteorological disasters in Canada were recorded in the global Emergency Events Database. These events included a winter windstorm (16 December 2007) that extended from southern Ontario through Quebec, killing four; extensive flooding in British Columbia in June 2007; and flooding through the spring of 2006 in British Columbia, the Northwest Territories, Ontario, Saskatchewan, and Newfoundland.

The impact of climate on society and ecosystems could change as a result of alterations in the physical climate system or in the vulnerability of society and ecosystems, even if the climate does not change (Kunkel et al. 1999b; Meehl et al. 2000a). 'Secure societies are those that have learned to live with their land as well as from it. Disaster reduction strategies will have succeeded when governments and citizens understand that a natural disaster is a failure of foresight and evidence of their own neglected responsibility rather than an act of God' (United Nations 2004).

Environment Canada (2003d) has compiled a list of the factors that make communities more or less vulnerable to natural hazards. See Table 1.1. It is estimated that weather-, climate-, and water-related hazards may account for nearly 90 per cent of all natural disasters. The findings of the Canadian Natural Hazards Assessment Project indicate that mitigating the risks of natural disasters in Canada requires more than advancements in science and technology. It is incumbent upon us, as a society, to be more aware of the concepts of risk and vulnerability and to consider these at all levels of decision-making. Canadians remain more vulnerable to natural disasters than they need be (Etkin et al. 2004). 'We're becoming more and more vulnerable to extreme weather. If the ice storm had happened 150 years ago, it would just have been a curiosity . . . We are so dependent on infrastructure because of increasing population and increasing wealth that we are becoming more and more vulnerable to severe weather disturbances' (McIntyre 2006).

How Is Severe Weather Defined and Classified?

Weather events can be classified as extreme according to various factors, such as the impact the event has economically (insurance costs), socially (loss of life), or environmentally (destruction of habitat) (Tompkins 2002). Extreme weather is a relative phenomenon; what may be considered extreme in one locale might be the norm in another. Weather events that are infrequent or rare are the ones that tend to be considered extreme or severe. The classification of an event often depends on what a region is used to experiencing and what it is prepared for. A 20 cm snowfall would be an extreme event for Washington, DC, for example, but not for Montreal. In

Table 1.1 Factors that make a community more or less vulnerable to a natural hazard.

Factors that make us **less** vulnerable:

- Better warning and emergency-response systems
- Greater economic capacity
- Well-established government disaster-assistance programs and private insurance companies
- Better government policies
- Community initiatives
- Social resilience
- Advances in science and engineering
- Major risk-reduction programs, such as the Red River Floodway

Factors that make us **more** vulnerable:

- Population growth (extending into hazard-prone areas)
- Urbanization
- Environmental degradation
- Urban expansion in hazard-prone areas
- Loss of community memory about hazardous events due to increased mobility
- An aging population (the number of Canadians over age 65 will increase to 1 in 5 by 2026, up from 1 in 20 in 1921)
- An aging infrastructure unable to cope with environmental loads
- Greater reliance on power, water, transportation, and communication systems
- Historical overreliance on technological solutions

Washington, such an event would come close to an emergency situation. In Montreal, it would merely be an inconvenience (Francis and Hengeveld 1998). Today's definition of severe weather is extremely broad. Given the concept of vulnerability, extreme weather can mean anything that causes hardship or has an economic impact (Street et al. 1997).

Extreme climate events and weather hazards can be more specifically defined, however. Extreme daily temperatures, extreme amounts of daily rainfall, unusually warm monthly temperatures over large areas, and storm events such as hurricanes, tornadoes, and hailstorms are but a few examples. Extreme events can also be defined by the impact they have on society. That impact may involve excessive loss of life, excessive economic or monetary loss, or both (Easterling et al. 2000).

Severe weather can also be categorized according to the spatial and temporal scale of the event (Figure 1.1). It can range from a tornado at the local scale, perhaps lasting only minutes, through to a snowstorm at the regional scale, lasting up to a day, through to climate change at the global scale, with a temporal period of 100 years and longer. It is important also to note that the time it takes for a severe weather event to occur (perhaps a few hours in the case of a tropical storm passing over a region) may be far surpassed by the amount of time it may take to recover from such an event (up to years or even decades in the most severe cases).

The Climate Severity Index (CSI) provides another way both to classify severe weather and to determine the locations where such extremes are likely to have the greatest impact (Phillips and Crowe 1984). The Canadian Institute for Climate Studies (2001) devised the CSI, including such indices as wind chill, length of winter, severity of winter, humidex, length of summer, warmth of summer, dampness, darkness,

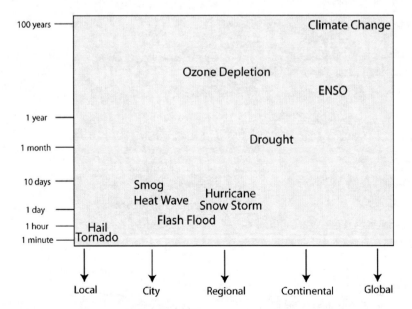

Figure 1.1 Spatial and temporal scales of weather and weather-related natural hazards.

Source: Modified from McBean and Henstra 2003.

sunshine, wet days, fog, strong winds, thunderstorms, blowing snow, snowfall, visibility, and freezing precipitation. Table 1.2 presents the indices and the percentage of each climate variable that constitutes the relative severity of one location as compared with another.

The intent of the CSI is to measure the impact of climate on human comfort and well-being and the risk of certain hazards to human health and life on a scale of 0 to 100, with a score of 100 being most severe. There are inherent issues with the CSI, the least

Table 1.2 Climate Severity Index (CSI) categories and subcategories.
Source: Modified from Phillips and Crowe 1984.

Sub-index	% of CSI	Description
Winter Discomfort Index	35	
Wind chill	15	Mean percentage of time in January that wind chill exceeds 1400 Wm^2
Length of winter	10	Number of months with mean daily temperature less than 0°C
Severity of winter	10	Mean daily temperature of coldest month
Summer Discomfort Index	15	
Humidex	5	Mean percentage of days with humidex greater than 30°C for an hour or more—highest 10-day value
Length of summer	2.5	Number of months with mean daily temperature of 10°C or greater
Warmth of summer	2.5	Mean daily temperature of warmest month
Dampness	5	Mean July wet-bulb depression
Psychological Index	20	
Darkness	7	Dependence of darkness on latitude
Sunshine	5	Mean annual number of hours with bright sunshine
Wet days	5	Mean annual number of days with measurable precipitation
Fog	3	Absolute frequency of hours with fog
Hazard Index	20	
Strong winds	6	Mean percentage frequency of wind equal to or greater than 30.6 km/hr—average of January and July
Thunderstorms	2	Absolute frequency of hours with thunder
Blowing snow	8	Absolute frequency of hours with blowing snow
Snowfall	4	Mean winter snowfall

Table 1.2 Continued		
Sub-index	% of CSI	Description
Outdoor Mobility Index	10	
Snowfall	2	Mean winter snowfall
Visibility	4	Absolute frequency of hours with fog, rain, or snow
Freezing precipitation	4	Absolute frequency of hours with freezing precipitation

of which being the values associated with the subcategories and the weight that different individuals might apply to such values. A Maritimer would likely argue, for example, that fog is not nearly as severe a condition as extreme summer heat. Nevertheless, when mapped, the index shows what might arguably be a relatively accurate representation of both the less severe and most severe locations in Canada (Figure 1.2). It is interesting to note the ways that the index can be manipulated to suit a particular purpose. The city of Lethbridge, Alberta, for example, reports a CSI of only 33 on its official website (www.chooselethbridge.ca), comparing this value with the relatively 'severe' indices of places like Winnipeg (51) and St John's (59) in an effort to draw people to its comparatively benign climate.

Weather and weather-related hazards vary in incidence across Canada. Tornadoes and hailstorms are most common on the Prairies and in southern Ontario; storm surges are common on the east coast; and winter storms occur across the country (McBean and Henstra 2003). Canada's least severe climates are found on southeastern Vancouver Island, in the Okanagan Valley of British Columbia, in southern Alberta, and in southern Ontario; its most severe climate is found in the Arctic Archipelago (Phillips and Crowe 1984). All these patterns will inevitably shift. Future climate changes may have a significant impact on current CSI trends. Since 1900, the mean annual temperature over southern Canada has increased by an average of 0.9°C (Bonsal et al. 2001). This can be translated into fewer days with extreme low temperatures during winter, spring, and summer and more days with extreme high temperatures during winter and spring. A less severe CSI ranking, however, can actually lead to severe impacts in some parts of the country. A lowering in CSI in northern regions might be associated with higher temperatures, leading to melting of former permafrost areas and eventually to the hazards associated with those conditions.

Environment Canada's weather watches, warnings, and advisories provide another useful way to categorize specific severe weather events (severe thunderstorm watch/warning; tornado watch/warning; freezing rain warning; heavy rain warning; frost warning; wind warning; marine wind warning; dust storm advisory; blizzard warning; heavy snowfall warning; winter storm warning; wind chill warning; and cold wave advisory). The methods that go into the forecasting and notification of such events are discussed in further detail in Chapter 2.

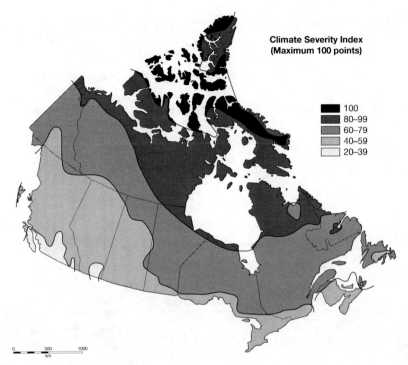

Figure 1.2 Spatial distribution of the Climate Severity Index.
Source: Modified from Phillips and Crowe 1984.

Introduction to Atmospheric Hazards in Canada

Approximately half of all Canadian disasters—whether natural or not—have been weather related, and this percentage has increased dramatically in recent years. The most expensive natural disasters this country has experienced have nearly all been weather related, with flooding being the main cause of the increase (Street et al. 1997; Environment Canada 2003d). Environment Canada's Meteorological Service has been working with a team of public and private sector partners to analyse trends in weather patterns. They have formed the Canadian Natural Hazards Assessment Project to determine explanations for trends and to develop ways to adapt to them better (Angas 2006). 'Canada has not been subject to the disastrous . . . cyclones, typhoons and floods which regularly take thousands of lives in countries like China and Bangladesh. Even the United States has more weather-related disasters because its larger population is subjected to far more hurricanes than Canada, and the USA is the most tornado-prone country in the world' (Jones 1992, 50). And yet Canadians should not be complacent, for we have had our share of weather-related disasters (Table 1.3).

Data on Canadian disasters show that although they are infrequent and difficult to predict, they are very expensive and can result in great personal loss. These studies reveal with a reasonable degree of certainty an increasing trend in the number and relative costs of disasters, even when adjusted for population growth (Dore 2003).

Table 1.3 Recent Canadian weather-related natural disasters.
Source: Modified from McBean and Henstra 2003.

Disaster	Number of Deaths	Economic Cost ($B Canadian)
Ice storm (1998)	28	$5.5
Edmonton tornado (1987)	27	$0.15
Barrie tornado (1985)	12	$0.2
Pine Lake tornado (2000)	12	$0.02
Saguenay flood (1996)	10	$1.5
Hurricane Juan (2003)	8	$0.15
Manitoba flood (1997)	4	$1
Calgary hailstorm (1991)		$0.36
BC blizzard (1996)		$0.2
Winnipeg flood (1993)		$0.16
Calgary hailstorm (1996)		$0.14

Table 1.4 Number of weather-related disasters and associated deaths and injuries in Canada, 1900–1999.
Source: Modified from Emergency Preparedness Canada 2002; Dore 2003.

Period	Number of Disasters	Number of Deaths	Number of Injuries
1900–9	7	303	5
1910–19	17	4077	353
1920–29	19	509	156
1930–39	15	1102	48
1940–49	21	499	398
1950–59	32	615	594
1960–69	49	452	30
1970–79	90	319	926
1980–89	107	903	1159
1990–99	130	1519	4336

Table 1.5 Canada's most expensive natural disasters.

*Billions of 1999 dollars; the range reflects the differing figures from various sources.
Source: Modified from Environment Canada 2003d; Etkin et al. 2004; Angas 2006.

Year	Type of Natural Disaster	Location	Cost*
2001–2	Drought	British Columbia, Prairies, Ontario, Quebec, Nova Scotia	5.0
1998	Ice storm	Ontario and Quebec	4.2–5.5
1979–80	Drought	Prairies	2.5
1988	Drought	Prairies	1.8–4.1
1984	Drought	Prairies	1.0–1.9
1996	Flood	Saguenay, QC	1–1.7
1950	Flood	Winnipeg, MB	1.1
1931–38	Drought	Prairie provinces	1.0
1991	Hailstorm	Calgary, AB	0.14–1.0

The number of weather-related disasters in Canada would seem to be on the rise (Tables 1.4 and 1.5), although this may in part be attributed to increasing vulnerability and not necessarily to increased frequency of severe weather events. It is interesting to note that even though the number of extreme events appears to be increasing, the number of deaths is variable for the decades of the twentieth century. With advances in health care, emergency response, and preparedness, the apparent increase in extreme events has not necessarily translated into more fatalities, although this could be an indication of more injuries that do not result in fatalities. To date, the most expensive natural disasters in Canada from 1900 to 2002 have been atmospheric (Figure 1.3).

Dore (2003) reports that a quarter of all hydro-meteorological disasters in Canada affect Ontario and Quebec (the most heavily populated parts of Canada). In recent years, however, Alberta has had a significant number of weather-related disasters as well. Reports such as these reflect economic impacts rather than personal or social impacts. A home destroyed in a remote location of Canada, assessed at a much lower value than a single home destroyed in downtown Toronto, will likely not be incorporated into such economically based reporting systems.

Figure 1.4 shows the general locations where costs associated with severe weather are greatest (as a measure of property and automobile losses due to hail, storms, tornadoes, flooding, wind, and snowstorms). It does not necessarily show the locations where the risk of a severe event is most likely, but rather the locations where the vulnerability of being impacted may be higher or lower. From Figure 1.5 it can be seen that more weather-related disasters have occurred in the Prairie provinces and in Ontario than have occurred on the west and east coasts and in northern regions of

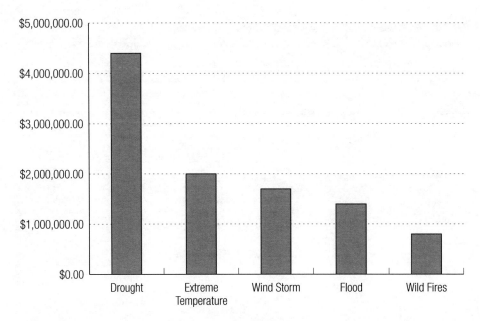

Figure 1.3 Economic damages by disaster type (US$ × 1000).
Source: Modified from E-M The OFDA/CRED International Disaster Database, www.emdat.net.
Université Catholique de Louvain-Brussels-Belgium

Canada. This would, in part, reflect the population distribution in the country as well as the impacts of agricultural droughts.

In recent years, many countries, Canada included, have experienced an alarming increase in natural disaster financial loss. Droughts are by far the most costly hazard, although they rank fourth in terms of frequency (Street et al. 1997). Prairie droughts have been the most costly, accounting for some $16 billion in losses in the past quarter-century. Since the early 1980s, four of the six most expensive natural disasters in Canada have been related to droughts. However, the 1998 ice storm was the single most expensive event for the Canadian insurance industry (Etkin et al. 2004). According to the Insurance Bureau of Canada, ever since the Edmonton tornado of 1987, the number of multi-million-dollar losses from weather disasters has been on the rise in Canada (before 1987, there was no single natural disaster with damages exceeding $1 billion anywhere in the world, let alone Canada) (Street et al. 1997). With the bulk of these disasters hitting since the late 1990s, the Canadian government has spent more than $13 billion to restore damaged infrastructure and uninsured properties (Angas 2006). 'The sociological and economic costs to Canadians from natural hazards are substantial, not only as a result of damages when events occur, but also due to adaptation and recovery. In particular, drought, flood and hail have had significant economic impacts. Understanding the impact these hazards have had can help us devise better policy tools to deal with them' (Etkin 1997). Researchers caution that the events we have experienced do not represent worst-case scenarios (Haque and Kilgour 2000). These summaries are of an economic variety and do not take relative impacts into consideration. Some disasters, although not high in economic losses, can nevertheless be catastrophic for the communities affected.

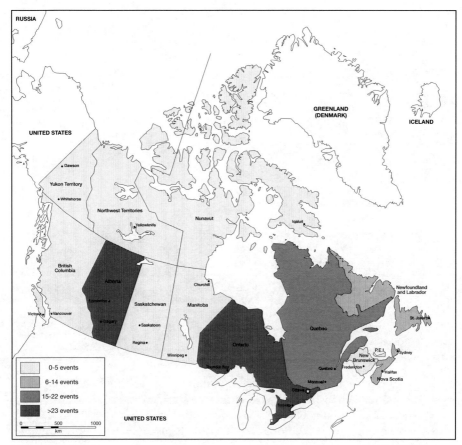

Figure 1.4 Weather-related natural disasters by province (major multiple-payment occurrences), 1987–2002.
Source: Modified from Insurance Bureau of Canada.

Canadian Disaster Mitigation, Preparedness, Response, and Recovery Activities

> *We have the knowledge and skills to make our communities safer. There are things we can do to reduce the potential for disaster, enhance preparedness for the disasters that do occur and improve our ability to respond and recover from them. We can modify behaviours and policies that increase our vulnerability to disasters.*
>
> Etkin et al. 2004

It is important to differentiate between the terms *mitigation, preparedness, response,* and *recovery*, as each involves a different type of action related to the threat of natural hazards. *Mitigation* involves long-term actions that reduce the risk of natural disasters (i.e., constructing dams and land use planning in high-risk areas). *Preparedness* involves planning for disasters and putting measures in place that will help communities and individuals cope in the event of a natural disaster (i.e., stockpiling essential goods,

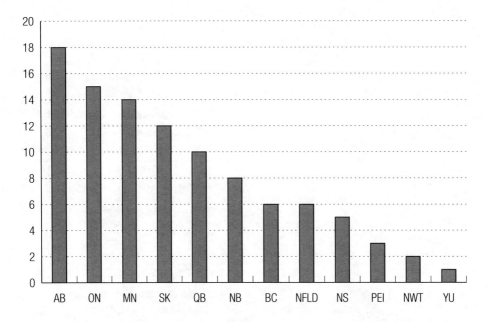

Figure 1.5 Frequency distribution of weather-related disasters in Canada by province, 1900–2000. Number of disasters = 473.

Source: Modified from Dore 2003.

preparing emergency plans). *Response* refers to the actions that would be taken in the event of a disaster (i.e., by police, firefighters, emergency medical personnel). *Recovery* involves actions taken to rebuild and restore communities to a pre-disaster state. There has been a tendency in the past for governments to focus on preparedness, response, and recovery rather than on mitigation. It is important, however, for communities to enhance their resilience through mitigation measures such as legislation, land use policies, engineering activities, warning systems, and public education programs.

The government of Canada supports a range of activities involving disaster risk reduction, preparedness, response, and recovery that are aimed at enhancing the capacity for disaster management (Foreign Affairs and International Trade Canada 2006). Public Safety Canada (PSC) is taking a leading role in the development of Canada's first-ever National Disaster Mitigation Strategy, which is aimed at reducing the risk and impact of natural disasters. PSC is modernizing Canada's emergency management legislation so that it might better reflect disaster mitigation as an integral component of comprehensive emergency management (mitigation, preparedness, response, and recovery). The PSC advocates a variety of measures to reduce the risk and impact of disasters, such as hazard mapping, land use zoning practices, hailstorm suppression, the burying of electrical cables to prevent ice buildup, diking, and disaster mitigation public awareness programs (PSC 2006b). PSC encourages Canadian natural hazards research, networking among emergency management stakeholders, and knowledge transfer through the Canadian Risk and Hazards Network (CRHNet), established in December 2003. The CRHNet will be registered as a partnership initiative under the United Nations Commission on Sustainable Development. Among the best-known examples of disaster mitigation in Canada is the Red River Floodway, constructed to protect the city of Winnipeg from the impact of flooding in the Red River Basin (see Chapter 5 for a discussion of this project).

In the event of a large-scale disaster, Canada provides support for disaster recovery through the Disaster Financial Assistance Arrangements (DFAA). When response and recovery costs exceed what individual provinces or territories can reasonably be expected to bear on their own, the Canadian government can use the DFAA as a means of helping provincial and territorial governments cover such expenditures. Since the program began in 1970, the government has paid out more than $1.4 billion in post-disaster assistance. Recent examples include payments for the 1997 Red River flood in Manitoba and the 1998 ice storm in Quebec and Ontario. Examples of eligible expenses include those for rescue operations, restoring public works and infrastructure to their pre-disaster condition, and replacing or repairing the basic, essential personal property of individuals, small businesses, and farmsteads (PSC 2006b).

The Canadian government is involved in two other initiatives related to disaster mitigation: Natural Resources Canada's Canadian Wildland Fire Information System (which provides daily and annual monitoring, mapping, and modelling of forest fire danger and activity) and the Meteorological Service of Canada's 24-hour weather watch and warning system (which alerts the public and mariners of impending severe weather).

Despite all the above measures, under normal circumstances mitigation is a relatively low priority for governments and individuals (Henstra and McBean 2004). Benefits of mitigation and disaster prevention are neither immediate nor tangible, and they are often expensive. It is hard to convince individuals and governments to make investments in the absence of imminent threats. The Insurance Bureau of Canada (2006) underscores the need for individuals and governments to be aware of risk and to consider ways to reduce vulnerability. The bureau encourages governments to embrace its three-part Natural Disaster Reduction Plan: (1) to invest in infrastructure, (2) to prevent recurrence, and (3) to develop a culture of mitigation. Planners and decision-makers should keep in mind that the losses associated with natural hazards are not inevitable. With appropriate planning to reduce vulnerability, the social and economic impacts of natural hazards on Canadians can be reduced. 'In a war with Mother Nature, we are going to lose. The best we can hope for is to minimize the losses through mitigation . . . this requires pro-active thinking and determined planning' (Downer 2007).

Link to Climate Change and Natural Climatic Oscillations

Over the twentieth century, the average temperature of the Earth increased 0.4°C–0.8°C. One of the anticipated effects of climate change is the possible increase in both frequency and intensity of extreme weather events, such as hurricanes, floods, and droughts. Two questions are now being asked: Are extreme weather events becoming more frequent? And if so, is this linked to climate change (e.g., Tompkins, 2002)? The focus of earlier climate-model studies of global warming focused mainly on changes in mean climate. It has been only since the early 1990s that researchers have begun to analyse climate models in order to predict possible changes in the weather and climate extremes of the future (Meehl et al. 2000b). With increased modelling capabilities and advances in computer technology, scientists are able to make predictions about the impacts of climate change in ways that they were never able to do before. The link between climate change and severe weather remains more contentious for some types of severe weather (e.g., hurricanes) than for others. Putting weather extremes in a climate-change context is challenging for several reasons. For one, extremes are rare by definition, which makes it difficult to draw statistical conclusions. For another, the most severe weather tends to affect areas smaller than global climate models can depict (Henson 2005). Yet it is clear that weather extremes are becoming an increasingly serious problem for our society, and there is also a reasonable probability that global warming will make the problem worse. The difficulty is that we don't know with certainty the extent to which the present wave of extremes is a natural climatic phenomenon; nor do we know the real potential for the intensification of extreme weather in a warmer climate. To diminish these uncertainties, we must devote much more scientific effort to the study of severe weather as a feature of climate. Even if we were not to face an anthropogenic climate change in the near future, we would still need to pay attention to measures that would help us adapt and respond to, as well as recover from, severe weather events. The fact that we might face more and perhaps stronger problematic weather systems in a climatically changed world underscores the need to expend additional effort on such measures.

Superimposed on the ongoing longer-term climate changes are shorter temporal scales of variability. In fact, variability is a natural feature of the climate system (Francis and Hengeveld 1998). The North Atlantic Oscillation (NAO, discussed in detail below) tends to switch phases every couple of years while also appearing to follow a longer cycle in which it is predominantly in one phase for 30–40 years and then predominantly in the other for the next 30–40 years. From 1900 until the late 1940s, the positive phase, in which pressure differences are large, predominated. From then until around 1980, the negative phase was more common. Since then, the oscillation has returned to its positive phase. The cause of the oscillation is not well understood, but it is clearly a natural phenomenon that affects the severity of winter weather in different parts of the North Atlantic region (Francis and Hengeveld 1998).

Elsewhere on the globe, El Niño–Southern Oscillations (ENSOs), by distorting atmospheric circulation patterns, bring profound changes to weather patterns in the tropics and even in the middle latitudes. Droughts in Australia and Africa, floods in Brazil and Paraguay, freak snowstorms in the Middle East, and monsoon rains in India and Indonesia have at various times been linked to ENSO conditions. In Canada, the effects of ENSOs vary considerably. ENSO and NAO in combination can have pronounced effects on weather in Atlantic Canada. An El Niño with a negative NAO can result in warmer and drier winter conditions. A La Niña combined with a positive NAO can result in more winter precipitation, which was certainly the scenario in the Atlantic provinces during the winter of 2007–8.

Scientific evidence suggests that because of climate change the frequency and intensity of severe weather-related events will continue to increase in the years to come. Why would greenhouse warming cause an increase in weather extremes? One reason is that the additional warming will change the distribution of heat and thus the flow of energy

The North Atlantic Oscillation

The North Atlantic Oscillation (NAO) is the term used to describe the oscillation of atmospheric masses over the North Atlantic region. The back and forth movement occurs between the so-called Icelandic low-pressure and the Azores high-pressure centres. The NAO is associated with climate variability from the North American eastern seaboard to Siberia and from the Arctic to the subtropical Atlantic (Stenseth et al. 2004). It achieves this large-scale climate-controlling capability by redistributing atmospheric mass between the Arctic and the subtropical Atlantic (Stenseth et al. 2004). The changes in atmospheric mass between the two areas causes changes in 'mean wind speed and direction over the Atlantic, the heat and moisture transport between the Atlantic and the neighbouring continents, and the intensity and number of storms, their paths, and their associated weather' (Stenseth et al. 2004).

Increased knowledge about the NAO and its cyclic manifestations has led to the creation of the NAO index. The index, which involves calculating the surface pressure difference between the Icelandic low and the Azores high, can reveal the general weather patterns found within the NAO's sphere of influence. There are two indices related to the two major phases of the NAO, the positive index and the negative index. The positive index is a period with a strong high-pressure centre in the Azores and a deep low-pressure centre located around Iceland. The implications of a positive index are stronger winter storms crossing the Atlantic Ocean on a more northerly track. During this period, Europe experiences warm and wet winters, northern Canada and Greenland see cold and dry winters, and the eastern United States sees mild and wet winters. During the negative index phase, there is a weak Icelandic low and a weak subtropical high in the Azores whose reduced pressure gradient results in weaker winter storms and an NAO path that is more east to west.

Sea surface temperatures (SSTs) are also influenced by the NAO. It has been documented that the strength of the NAO directly affects SST. When there is a positive NAO, the trend is towards colder water arriving in the sub-polar North Atlantic. Studies have also shown that air-to-sea heat exchanges and surface winds associated the NAO are related to SST anomalies (Stenseth et al. 2004).

Research has demonstrated that these oscillations can last 30 years or more. Recent studies have highlighted that the oscillations have hemisphere-wide effects and that the NAO is linked to changes in the Arctic Ocean. When the NAO is in a warm phase, Atlantic water can travel farther into the Arctic. As a result, the sea ice thins. During a cold phase, lower-level winds strengthen, isolating the Arctic Ocean from warmer, saltier water to the south and allowing the sea ice to gain in thickness (Aguado and Burt 2007).

Source: Aguado, Edward and James Burt. *Understanding Weather and Climate*, 4th edn., © 2007, pp. 249, 257. Reprinted by permission of Pearson Education Inc., Upper Saddle River, N.J.

through the climate system. This will in turn alter the circulation patterns of the atmosphere and oceans, and it will also modify the hydrological cycle by which water is circulated between the Earth's surface and the air. As a result, the position of many of the world's major storm tracks could shift significantly. Some areas would be exposed to more storms and heavier rainfalls, while others might see formerly reliable rainfalls give way to prolonged dry spells (Francis and Hengeveld 1998). A second and more compelling reason for suspecting a link between greenhouse warming and weather extremes is related to the potential effects of a warmer climate on the physical processes that generate different types of weather events. A virtually certain outcome of a rise in temperature is a widespread increase in the amount of water that is moved through the cycle. This is because higher temperatures not only increase evaporation and transpiration but also raise the air's capacity to hold moisture. Consequently, more moisture will be available to fall as rain or snow. Add to this a more unstable atmosphere due to increased convection over warmer land and sea surfaces and the result is an increased potential for major precipitation events in many parts of the world. Because of changes in large-scale circulation patterns, as well as regional differences in hydrological processes, the resulting increase in precipitation will not be spread uniformly around the world (Francis and Hengeveld 1998). In 2005, for example, regionally drier than average conditions were widespread across eastern Australia, parts of western Europe, and the United States. Rainfall was well below average in the Amazon Basin, producing the worst drought in 60 years in some parts of the region and resulting in the lowest water levels in at least 30 years along the world's second-longest river. Across southeastern Africa, long-term drought affected parts of Mozambique, southern Malawi, and Zimbabwe. In contrast, several regions of the globe experienced heavy precipitation during 2005. On 26 July 2005, a new 24-hour rainfall event record was established in the city of Mumbai when over 944 mm of rain fell. In Colombia, heavy rains in October and November triggered floods and landslides. In Saudi Arabia, heavy January rains produced some of the worst flooding in 20 years in the city of Medina. In August, heavy rainfall affected areas of central and eastern Europe, with flooding reported in sections of Romania, Hungary, and Macedonia. Farther to the west, flooding also affected areas of Germany, Austria, and Switzerland (NCDC 2005).

By some accounts, Canadian cities have already begun to experience the changing hazards associated with climate change (McBean and Henstra 2003). One irony is that the future climate of Canada, which will become warmer with increasing greenhouse gases in the atmosphere, will likely produce less personal discomfort for Canadians (Table 1.6). The bad news is that the climate could become more hazardous owing to an increasing frequency of severe weather events. Climatologists are becoming increasingly concerned that the volatile weather experienced in 1996 (the year that included the Saguenay River flood, a severe hailstorm in Alberta, and a damaging blizzard in British Columbia) might be a dry run for the extreme conditions we might expect from a warming climate (Street et al. 1997).

The 2007 report of the Intergovernmental Panel on Climate Change (IPCC 2007b) tells us that there is a greater than 99 per cent chance that in the future the Earth will have warmer temperatures and more frequent hot days and nights. This represents a change from the 90–99 per cent level of certainty for the same phenomena reported in the 2001 IPCC report. With 90–99 per cent certainty, there would be an increase in heat

Table 1.6 Likelihood of future climatological trends.
Source: Modified from IPCC 2007b.

Phenomenon and Direction of Trend	Likelihood of Future Trends Based on Projections for 21st Century
Warmer and fewer cold days and nights over most land areas	Virtually certain
Warmer and more frequent hot days and nights over most land areas	Virtually certain
Warmer spells/heat waves. Frequency increases over most land areas.	Very likely
Heavy precipitation events. Frequency (or proportion of total rainfall from heavy falls) increases over most areas.	Very likely
Area affected by drought increases.	Likely
Intense tropical cyclone activity increases.	Likely
Increased incidence of extreme high sea level (excludes tsunamis).	Likely

Judgmental estimates of confidence:

- Virtually certain: Greater than 99% chance that a result is true
- Very likely: 90–99% chance
- Likely: 66–90% chance
- Medium likelihood: 33–66% chance
- Unlikely: 10–33% chance
- Very unlikely: (1–10% chance)
- Exceptionally unlikely: Less than 1% chance

waves over most land areas and heavy precipitation events would increase. Subsequent chapters in this text will illustrate what the global climate-change scenarios and implications might be for Canada. Chapter 2 will set the stage by describing how severe and extreme weather is forecast and how Canadians are notified of such events.

Chapter Summary

Natural disasters occur when a hazard triggers vulnerability and the damage is so extensive that the affected community cannot recover through the use of its own resources. Natural hazards can affect anyone, anywhere. People are threatened by hazards because of their social, economic, and environmental vulnerability. Approximately half of all Canadian disasters—whether natural or not—have been weather related, and this percentage has increased dramatically in recent years. There has been a tendency in the past for governments to focus on preparedness, response, and recovery. It is also important, however, for communities to enhance their resilience through mitigation measures such as legislation, land use policies, engineering activities, warning systems, and public education programs. One of the anticipated effects of climate change is the possible increase in both the frequency and the intensity of extreme weather events such as hurricanes, floods, and droughts. Two questions must now be asked: Are extreme weather events becoming more frequent? And if so, is this linked to climate change? Scientific evidence suggests that because of climate change the frequency and intensity of severe weather-related events will continue to increase in the years to come.

Chapter Two

Severe Weather Forecasting

Objectives

- To review historical and contemporary weather forecasting in Canada
- To explain data sources and weather modelling for purposes of weather forecasting
- To explain the use of radar and satellite technology in severe weather forecasting
- To review how and when severe weather warnings are issued in Canada
- To introduce the concept of a 'citizen forecast'

Introduction

The weather forecast is of considerable importance to the average Canadian, whether for reasons relating to recreational activities and mere curiosity or to work and safety. We curse forecasters for inaccurate predictions but rarely praise their accomplishments. We have never, however, held Canadian forecasters accountable for their inaccuracies, as was the case in Brazil in 2001. A veteran local meteorologist in Rio de Janeiro predicted rain on New Year's Eve, prompting many people to stay at home. The cold front that was to have brought the heavy rain broke up earlier than expected, and so the rain didn't come. This might sound like a familiar story to many Canadians, as we expect forecasts to be incorrect at least some of the time, but the story doesn't end there. The mayor of Rio decided that the forecaster should be held accountable for the poor turnout at the city's annual New Year's celebration, and so the forecaster was sentenced to six months in prison (de Blij et al. 2005). We might not go so far as to expect prison terms for bad forecasts, but Canadians do expect a certain degree of accuracy, especially when it comes to severe weather. The prediction of precipitation amount is arguably the greatest challenge facing the weather forecaster today. Moreover, this is a weather element of particular interest to the general public (Peel and Wilson 2006).

Canadians are probably more aware of weather conditions today than they have been in the past. We rush to the stores to stock up on supplies when severe weather is predicted and make plans for our daily activities according to the forecast for the day. This is due, in no small part, to the improvement in the news media's weather coverage in recent years. The first weather forecaster to appear on Canadian television was the meteorologist Percy Saltzman. His main props were chalk (famously tossed in the air at each broadcast's conclusion) and a map of Canada on a board. Today, weather reporters use animated maps, satellite imagery, and colour radar. Moreover, with millions of Canadian households now wired to receive the Weather Network on cable TV and websites such as wunderground.com and weather.com via computer, viewers and Internet surfers are better informed than ever.

North America has its share of volatile weather conditions, including the highest surface wind speed (de Blij et al. 2005). There are many more weather extremes in North America than in other continents because of the interaction between the polar and tropical air masses in North America and the irregular pathways of the polar front jet stream. Within North America, the northeast is a region of particularly changeable weather, and thus more forecasts have gone wrong for this region than for any other part of the continent. Winter storms have proven especially difficult to predict, even a few hours ahead, and the need for better snow and ice warnings in this heavily urbanized region has prompted the intensification of research efforts.

Because Environment Canada generally makes predictions for broad regions and these predictions are unreliable to some degree, the needs of businesses whose activities are closely related to weather conditions have given rise to a private forecasting industry. Private forecasters are far more common in the United States than in Canada, but their use is growing in this country and Canadians are likely most familiar with them through the news media. The Weather Network's patented Pelmorex Forecast Engine (PFE) technology, for example, allows meteorologists to issue weather forecasts for the entire country down to a 10 km^2 grid, although this in no way implies great accuracy.

Forecasts, in general, can be issued up to five days in advance. Meteorologists are experimenting with 10–14-day predictions, but these are too unreliable to be issued to the public (Environment Canada 2006d). The Weather Network and some commercial weather forecasters do issue forecasts of longer trends. The Weather Network issues a 14-day trend line that illustrates expected above and below average temperatures as well as expectations for precipitation or sunny weather. Forecasters continue to use the synoptic scale and mesoscale analysis techniques that were developed in the pre-computer age, however. This chapter will explain more recent advances in forecasting tools and how they serve weather forecasters.

Weather Forecast Quality

Weather forecasts tend to elicit a sceptical response from the public. Accurate or close to accurate forecasts can easily be forgotten with one 'bad' forecast. Moreover, severe weather can be especially challenging to forecast given the low statistical frequency of such events, the wide variation in model predictions, and the public's heightened awareness and attention to the details of the forecast. Studies of weather forecast quality and/or value have received considerable attention in the academic literature in recent years (e.g., Richardson 2000; Stephenson 2000; Thornes and Stephenson

2001). The quality of a weather forecast depends in large part on its reliability, accuracy, and the skill of the forecaster. Some researchers will argue that forecasts remain a game of pure chance, but one in which a skilled forecaster can increase accuracy through an expert analysis of meteorological information. Stewart and Lusk (1994) report that the performance of a forecaster depends on three factors: the environment about which the forecasts are being made (how predictable is it?), the information system that brings data about the environment to the forecaster (how reliable is the information provided?), and the forecaster's cognitive skills (the forecaster's perceptual and judgmental processes). Notwithstanding such cautions, the fact remains that weather forecasts and the public's access to weather-related information have improved substantially from the earliest days of forecasting over a century ago.

The Way It Was

In the early years of the Canadian Meteorological Service, a storm warning was defined as 'a publication of an opinion to the effect that shortly after a time specified or implied, a storm will probably occur in some portion of a certain region within a radius of 100 miles of the port warned' (Thomas 1971). In 1874, there were 35 storm-warning stations in eastern Canada, and on 56 individual days during that year, storm warnings were dispatched to the proper towns. By 1879, public forecasts were being telegraphed from Ontario to 125 places. Local recipients were expected to post them on public bulletin boards outside the local telegraph office, post office, railway station, or school. Forecasts were also sent to newspapers for publication in the evening editions.

An interesting innovation of the early 1880s was the dissemination of weather predictions by means of display discs attached to railway cars. The signal word to be displayed was telegraphed each day to the railway agents, who would change the signs on the cars each morning in an attempt to provide a reliable weather prediction service for the farming community along the railway line between Windsor, Ontario, and Halifax, Nova Scotia. The symbol on the disc would inform the farmers working in fields along the tracks of the expected weather conditions: a full moon meant sunny skies; a crescent moon, showers; a star, prolonged rains. However, through neglect, the local train hands did not always keep the signal discs up to date, and thus the arrangement was dropped after a decade or so (Weather Doctor Almanac 1999).

An annual report of the Canadian Meteorological Service in the early 1890s mentions that weather forecasts were requested by many people whose jobs were affected by changes in the weather—shippers, brewers, fishmongers, and proprietors of skating rinks. Railway companies, too, were asking for and receiving special forecasts of winter snowstorms and thaws. In 1889, Canada's Department of Agriculture began to cooperate with the Canadian Meteorological Service, charging personnel at four new experimental farms with the task of observing and reporting the weather.

At 35 Canadian ports and harbours along the Great Lakes, St Lawrence River, and Atlantic coast, personnel posted wind warnings by hoisting either wicker baskets, cones, or drums up a mast or pole, the type of object indicating the expected strength of the approaching storm—an improvement over the American practice at that time of only indicating that winds would be strong. Later, lanterns were used so that warnings could also be sent by night. The wicker basket and drum signals reportedly flew until the 1950s, when the last storm station was decommissioned (Weather Doctor Almanac 1999).

By the early 1920s the availability of weather forecasts in daily newspapers was an accepted part of Canadian life. Forecasts were issued twice daily, at 10 a.m. and 10 p.m., each for the ensuing 36 hours, based on data from about 36 Canadian, 5 Newfoundland, 1 Bermudian, and 100 American stations. In addition to the forecasts issued in the press and received at telegraph offices, the government's new wireless stations had begun broadcasting forecasts for the benefit of shipping on the Great Lakes and along the Atlantic coast.

With the outbreak of war in the Pacific in December 1941, the broadcasting of weather information and forecasts was banned by both the Canadian and American governments, although brief district forecasts were published in the local press in central Canada. The concern was that the forecasts could potentially assist the enemy. Even baseball announcers were prevented from commenting on the weather. On one occasion when play was suspended due to rain, the announcer reportedly told listeners to 'stick your head out of the window if you don't understand the reason for the suspension' (Environment Canada 2005e). It was not until the end of the war in Europe that Canadian newspapers were again permitted to publish any weather information they wished and radio stations could broadcast reports and forecasts.

Until the 1950s, weather forecasters depended on their own experience and ability when they attempted to interpret current and recent weather conditions. Meteorologists would largely base their forecast on a comparison of current weather conditions with similar conditions they had experienced previously. This approach led to the development of several 'rules of thumb' (or the 'analogue approach'). Although this method may have worked well enough for typical weather patterns, it was not particularly useful for severe or extreme weather conditions, because, by their very nature, these conditions are less frequent and therefore offer fewer historical examples against which to make comparisons. In the last few decades, the experiential method has given way to numerical weather forecasting (discussed below), which has come to occupy a dominant position in the field of weather forecasting.

The Forecasting Procedure

Broadly speaking, weather analysis and forecasting involve a succession of tasks (WMO 2004):

- Understand present weather conditions (weather analysis) utilizing technology such as satellite imagery, water vapour imagery, and model analyses from the last 24 hours.
- Obtain the pertinent information from at least one numerical model to assess the future evolution of the atmosphere in order to determine the most likely scenario (synoptic weather forecasting). This information may be compared to observations at the surface and at uppers levels, especially in the sensitive regions of the atmosphere (e.g., baroclinic zones).

- Predict the consequences of the expected synoptic situation in terms of weather elements (weather elements forecasting), evaluate the risk of the occurrence of hazardous phenomena (risk assessment), and synthesize the results of the analysis on a graphic document.
- Prepare the meteorological information (including possible warnings) to be issued to the public.

Data Acquisition

The starting point for any weather forecast is the current weather conditions. Even when making forecasts for a small geographic area, a forecaster needs data from a much larger region. The World Meteorological Organization (WMO), under the auspices of the United Nations, thus plays a very important role in overseeing the collection of weather data from its 188 member nations. Every six hours, beginning at midnight Greenwich Mean Time (GMT), also known as Universal Coordinated Time (UTC), sets of standard observations of the local state of the atmosphere are taken around the globe by about 10,000 land-based stations, more than 1000 upper-air observation stations, and at least 9000 ships, buoys, aircraft, and satellites in transit, and reported to the meteorological services of individual nations. High-powered computers then rapidly record, manipulate, and assemble these data to produce synoptic weather charts. The member nations of the WMO maintain their own meteorological agencies that obtain and process the data and issue regional and national forecasts. In Canada, the Canadian Meteorological Centre (CMC) of the present-day Meteorological Service of Canada (MSC) fulfils this function.

The Meteorological Service of Canada was established in 1871 as a national program for the official recording and observation of climate in Canada. Today, Environment Canada oversees a number of divisions that relate to weather and climate, including the CMC and the MSC. Environment Canada's weather forecasting services include a weather-warning service and five-day forecasts that are available 24 hours a day, 365 days a year. Environment Canada oversees six regional weather centres across the country; these centres provide Canadians with toll-free and pay-telephone services that cover their own and surrounding communities. Environment Canada's Weatheradio transmitters, located across southern Canada, reach 92 percent of Canadians, and Weatheradio forecasts for inland and coastal waters, as well as for the Arctic, are also transmitted.

Within Canada, there are more than 2000 observation stations that report surface weather conditions (Figure 2.1). Each station obtains readings from a number of instruments, such as thermometers, barometers, rain gauges, hygrometers (to measure the moisture of the air), weather vanes, and anemometers. Surface conditions are obtained from this network, but this information is supplemented with vertical details of the atmospheric conditions gained from radiosonde data. Radiosondes are radio-equipped instrument packages that are carried aloft by balloon. Canada's 37 radiosonde stations are part of a worldwide network that transmits readings to receiving facilities on the ground every 12 hours (noon and midnight GMT).

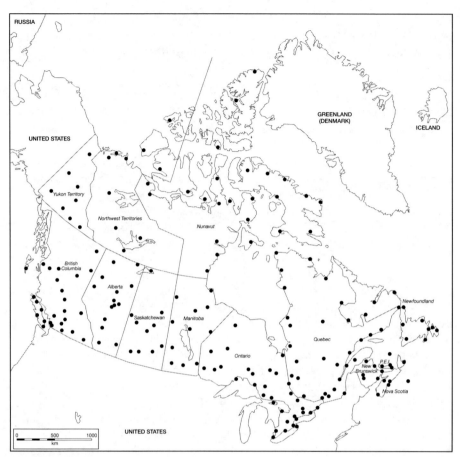

Figure 2.1 Location of Meteorological Service of Canada weather stations.

Source: Modified from Natural Resources Canada, *National Atlas of Canada* (2007).
Available at http://atlas.nrcan.gc.ca/sitefrancais/english/maps/index.html.

Weather Forecasting Models

Observations are crucial to weather forecasting. Many thousands are received each day, and these are processed, quality controlled, and monitored. After observations have been processed and quality controlled, they need to be incorporated, or assimilated, into the numerical model. In carrying out this very important process, the forecaster obtains a representation of the current state of the atmosphere—that is, an analysis—from which a forecast is obtained. The forecaster may then analyse satellite imagery and, following this, typically, work with an ensemble weather-forecasting model.

Weather and environmental prediction computer models provide weather forecasters with projections of the future state of the atmosphere from which they can discern the tell-tale signs of severe weather systems as well as when and where they might occur. Weather models are very useful in forecasting what the weather will be in a given time period (from 3 hours to 10 days). Although models have biases and sometimes miss weather systems, they are among our best tools for predicting the weather.

Numerical weather prediction (NWP) is a staple in the forecaster's toolkit. NWP is based on projections taken in small increments of time. For example, a forecast is prepared for the weather 15 minutes from now. Once those conditions are determined, the computer repeats the process by using them to make predictions for 15 minutes after that, and so on until the desired future time of the forecast is reached. Unavoidable and inevitable errors begin to creep in immediately, and these are magnified as the length of the forecast period increases. NWPs provide forecasts of temperature, pressure, moisture, rainfall, and winds. Forecasters examine the features predicted by the model to come up with a forecast for that day or for the next several days.

The various weather agencies around the world develop their own numerical models and typically maintain a suite of models rather than a single program. The models are updated continually, and from time to time new models are introduced and older ones retired. Although there are significant differences among them, all numerical models follow the same general procedure:

- *Analysis phase:* Observations are used to supply values corresponding to the starting (current) state of the atmosphere for all the variables carried in the model.
- *Prediction phase:* Beginning with values delivered by the analysis phase, the model uses equations to obtain new values for a few minutes into the future. The process is repeated over and over until the end of the forecast period is reached (e.g., 24 hours).
- *Post-processing phase:* A series of maps are produced depicting sea-level pressure, 850 mb (millibar) heights and temperatures, 700 mb heights and vertical velocities, 500 mb heights and absolute vorticity values, and precipitation amounts.

The upper-level maps that are analysed in the post-processing phase provide forecasters with information such as maximum expected temperatures, locations of frontal systems, and the trend in the jet stream. With an average sea-level pressure of approximately 1013 mb, the 850 mb maps indicate the heights above the surface at which point that pressure would be met, and so on through to the highest-level 250 mb maps where the identification of the jet stream can best be seen. After studying the maps produced in the post-processing phase, forecasters often create a manual forecast from the NWPs, supplemented by other analyses and tools.

Although models are extremely useful to forecasters, they are flawed in many ways. The equations used by the models to simulate the atmosphere are not precise, and there are many gaps in the initial data from areas in the mountains or over the ocean. To predict the weather for even a few days in advance, the forecaster must know the current state of the atmosphere in order to have the starting conditions for the numerical weather prediction model. In practice, weather observations on land and sea are insufficient for an unambiguous representation of the atmosphere, as, in addition to the errors one might expect in any set of measurements, the geographic and time distribution of observations is uneven, with some areas and levels in the atmosphere covered poorly or not at all. Bringing the observational and forecast data together in a mathematically rigorous fashion and ensuring that the component fields are in physical balance comprise a process known as *data assimilation*. Estimating the state of the atmosphere on the basis of such data is called the *analysis*. Overcoming data-assimilation problems is paramount in the quest to achieve more accurate short- and long-term forecasts. Research on computer models is ongoing and is focused on higher-resolution models, ensemble model forecasts, and improved methods for the incorporation of observations (National Severe Storms Laboratory 2006).

The Canadian 'Ensemble Prediction System'

During the past decade, ensemble forecasts have become the preferred operational tool for generating probabilistic forecasts of future weather events. Consequently, tools for assessing and comparing ensemble forecasts and more generally probabilistic forecasting techniques are in high demand. The first Meteorological Service of Canada global ensemble prediction system (EPS) became operational at the Canadian Meteorological Centre in January 1998 (Houtekamer et al. 1996). This system is currently undergoing major research, investigations, and verifications that may lead to the gain of about one day in the temperature forecast. Further work is being done on forecast systems that support forecasters and aid in decision-making (Toth et al. 2005).

The Canadian ensemble system is based on a 'system simulation experiment' (SSE). In an SSE, it is considered that all elements of the forecast system, including the model, the observations, and the analysis, are subject to uncertainty. The elements of the system are perturbed in different ways for different elements of the ensemble. In addition to considering uncertainty in both the analysis and the model, the Canadian system, in its current 16-member configuration, offers a truly multi-model ensemble: 'We use two completely different models, the older global spectral model (SEF) and the newer global version of the generalized environmental multiscale (GEM) model' (Wilson 2001).

The usefulness of an ensemble prediction system resides mostly in the variety of possible solutions to a given meteorological forecast problem that the system can offer. The differences between all weather scenarios presented by each member of the EPS (or the variance within the ensemble forecasts) lead towards the study of the spread-skill relationship. If such a relationship exists, it would be possible to associate higher skill with greater confidence in forecasts where the ensemble variance is low, and vice versa (Verret et al. 2006).

The North American ensemble forecast system is a new weather modelling system run jointly by the Meteorological Service of Canada and the US National Weather Service (NWS). The system provides numerical weather prediction products to weather forecasters in both countries for a forecast period of two weeks into the future. The National Meteorological Service of Mexico also participates, resulting in a North American-wide weather-forecasting capability.

Radar Technology

Weather radar is a type of radar used to locate precipitation, calculate its motion, estimate its type, and forecast its future position and intensity. Radar (a short form for **ra**dio **d**etection **a**nd **r**anging technology) was developed just prior to World War II as a method of detecting and locating hostile aircraft at long distances. At first, storms were considered a nuisance, for they obscured valuable data. However, as users became familiar with the technology, they came to recognize the value of radar in detecting weather, particularly storms. Weather radar is now invaluable in the forecasting of possible tornadoes (using the so-called hook echo signature on the radar image) as well as in picking up squall-line thunderstorms. Modern weather radar is mostly Doppler radar, which is capable of detecting the motion of rain droplets in addition to the intensity of the precipitation. Radar operates by sending out a beam of energy (radio waves) and then measuring both how much of that beam is reflected back by the objects (e.g., rain, snow, and sleet) that receive it, and how long it takes for the reflected beam to return. The more of the beam that is sent back, the higher the reflectivity of the object, and this is indicated by the brightness of the colours on the radar image. Reflectivity is a function of precipitation intensity and variety. Ice surfaces (such as hail and sleet) easily reflect energy, causing light sleet to appear heavy on the radar image. Snow can scatter the beam, with the result that heavy snow appears light on the image. Radar images are colour-coded to indicate the intensity of the precipitation.

Doppler weather radar facilitates the detection of high-impact weather and enables forecasters to determine the severity of a weather event with much greater precision than would otherwise be possible. This technology is used by the Meteorological Service of Canada to enhance the quick release and accuracy of forecasts of severe weather events, such as blizzards, tornadoes, freezing rain, thunderstorms, strong winds, hurricanes, hailstorms, and heavy snow or rain (Environment Canada 2004h). Doppler weather radar helps meteorologists determine motion and structure within weather systems. While conventional weather radar indicates the location and intensity of precipitation, Doppler processing adds detailed information on the movement of precipitation within these meteorological systems. Instead of measuring the speed of one object, Doppler weather radar, using microwave energy, measures the motion of millions of snow or water droplets in a storm system. The physical characteristics of the wave are noted when it is transmitted and again when it returns after bouncing off the droplets. Through a complex process, the change or shift in frequency of the returned signal is used to determine the motion of the droplets (MSC 2002). Waves reflected by a system (e.g., precipitation) that is moving away from the receiving antenna change to a lower frequency, while waves from a system that is moving towards the antenna change to a higher frequency. The computer technology associated with the Doppler system uses the frequency changes to show the direction and speed of the winds blowing around the precipitation. These radar images help forecasters more clearly understand what will happen through the duration of a weather event.

Experience in the United States and Canada has demonstrated that Doppler weather radar can be a very useful tool in improving the prediction of severe weather events. Meteorologists using Doppler weather radar along with other state-of-the-art tools and techniques are able to

- identify and more precisely define the areas where severe weather is likely to occur; and
- during severe weather events in the summer, search for characteristic patterns indicating a high probability of severe thunderstorms, hail, and tornadoes.

Figure 2.2 Location of Doppler weather radar stations covering Canada.
Source: Modified from http://badc.nerc.ac.uk/community/poster_heaven_old/Poster07E.pdf.

The goal is to be able to provide sufficient warning to the public to ensure their safety and security and to assist in the implementation of emergency planning with a view to minimizing damage and economic loss (National Radar Program 2005).

The area of coverage of the Doppler weather radar network across Canada was determined principally by criteria concerning population density and the probability of the occurrence of severe weather (Figure 2.2). The network provides virtually continuous coverage over the populated areas of the country most susceptible to severe weather. Of the 31 Doppler radar stations in the Canadian network, the Marion Bridge station in Nova Scotia is one of the more recent, put into service in 2002 (Figure 2.3). Located in the community of Marion Bridge in Cape Breton, the Marion Bridge Doppler radar was the first Canadian Doppler weather radar to capture a 'landfalling' hurricane. Hurricane Gustav made landfall as a category 1 storm along the southern Cape Breton coast, near St Esprit Island, during the early hours of the morning on 12 September 2002. This storm brought heavy rain (up to 102 mm), high

Figure 2.3 Marion Bridge Radar Station

winds (up to 122 km/hr), and minor flooding from a storm surge around Atlantic Canada (CHC 2002a).

The general public may be familiar with Doppler radar imagery through the Internet or through television weather broadcasts. These data can be included as part of any service agreement that the television stations may have with Environment Canada. Environment Canada considers the provision of Doppler radar imagery and data a specialized service and therefore subject to a fee to recover costs. Some television stations have their own Doppler weather radar.

Satellite Imagery

Today, the dozens of orbiting weather satellites that constantly monitor the Earth are perhaps the most important source of three-dimensional atmospheric data (see photo of Hurricane Bob in colour insert section). Many of these satellites operate within longitudinal polar orbits about 800 km above the Earth, surveying a different meridional segment of the surface during each revolution. In this manner, a complete picture of the

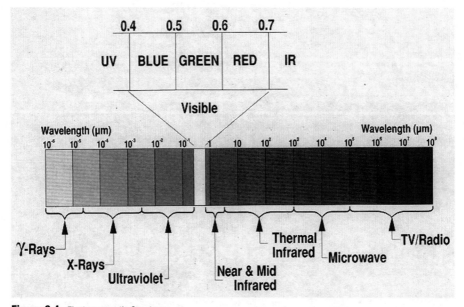

Figure 2.4 Electromagnetic Spectrum
Source: Modified from http://casde.unl.edu/tutorial/rs-intro/images/electsp.gif
Consortium for the Application of Space Data to Education (CASDE) project, University of Nebraska-Lincoln. Funding provided by NASA.

globe can be assembled every few hours. More and more satellites are now being placed in much higher orbits (35,800 km) above the Equator, called geosynchronous orbits—that is, they revolve at the speed that the planet rotates and remain stationary above a given surface location. A 'fixed' satellite at this altitude can monitor the same one-third or so of the Earth continuously, and with infrared (IR) capability, it can perform that task in darkness as well as in daylight (de Blij et al. 2005). The IR capacity also allows the satellite to detect the temperature of the clouds; the brighter colours on the IR images indicate higher and stronger storms (Figure 2.4). These satellites, particularly the Geostationary Operational Environmental Satellites (GOES), are critical tools for both forecasters and researchers. Orbiting the Earth at speeds matching the Earth's rotation at a height of 35,800 km, they provide measurements of cloud cover, both visible and infrared, and water vapour. Visible satellite images, unlike the IR images, are actual photographs of the Earth. These images show low clouds, fog, and snow cover in greater detail than IR images can. Ever more sophisticated sensing equipment—for example, microwave interferometers, which help detect surface marine winds—is continuously being deployed on geostationary platforms (Bowyer 2007).

Recent research is focusing on integrating satellite data into a multi-sensor approach to the severe storm-warning process. With radar and satellites combined, forecasters can have a better look at how storms are evolving and which ones may become severe. Canada uses two US satellites to assist in its identification of severe weather. Each satellite provides both visible and infrared images four times an hour. Visible images are available only in daylight hours, but the infrared images are available 24 hours a day. During severe weather, the satellites provide images at five-minute intervals, which permits better tracking of storms.

The Canadian Lightning Detection Network

The Canadian Lightning Detection Network (CLDN) was deployed in 1998 to increase public safety by allowing meteorologists to detect and monitor thunderstorms at an early stage in their development. Thunderstorms are always accompanied by lightning and may produce dangerous and damaging weather. In Canada, lightning kills about seven people and seriously injures 60 to 70 each year (Environment Canada 2008b).

The CLDN is made up of 81 remote lightning sensors that stretch across Canada, from Yukon to Sable Island, Nova Scotia (Figure 2.5); the sensors are owned by Environment Canada and operated by Global Atmospherics, Inc., in Tucson, Arizona. The primary design objective for the sensors was that they have a detection efficiency of 90 per cent–plus and accuracy of 500 m or better over those areas of Canada that have five thunderstorm days a year or more. In addition to the achievement of the primary objective, the Meteorological Service of Canada had to ensure that all major population centres, forested areas, and electricity production and transmission facilities—as well as all major aviation corridors and areas—were adequately monitored (MSC 2005).

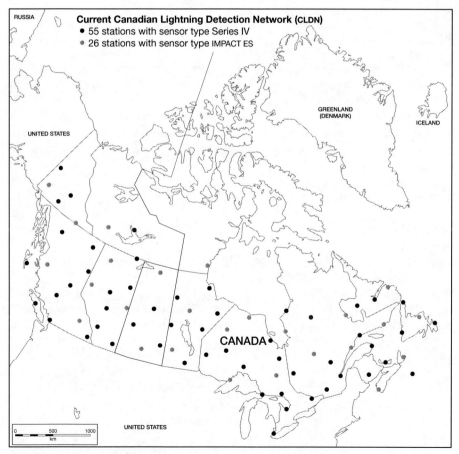

Figure 2.5 Location of Canadian Lightning Detection Network (CLDN) sensors.
Source: Modified from Environment Canada 2008b.

Individual sensors pick up information about lightning events such as the location relative to the sensor, the time of the event (using the Global Positioning System [GPS] time), and the amplitude and polarity. The sensors transmit this information via satellite to a TeleSat Canada hub in Toronto and hence by land line to Tucson, where the data from the 81 sensors in Canada are combined with the data from the 106 sensors in the United States. The combining of the data gives users in each country the benefit of the sensors in the other, so that both countries receive very high quality data. The computers in Tucson take the data from multiple sensors to compute the actual location of the lightning event, the time it occurred, its amplitude and polarity, and other data relating to the characteristics of the event. These data are then sent via satellite to the real-time data users and are usually displayed on users' screens within about 20–30 seconds of the event. Users include, among others, forestry organizations (for the early detection of forest fires), electric power companies (for storm tracking and the forensic analysis of lightning-caused outages and damage), aviation personnel (for ground crew safety and aircraft routing), and insurance companies (to verify claims for lightning-caused damage).

Seasonal Forecasts

Seasonal forecasts have been receiving more and more attention as researchers and the general public have become interested in knowing what the weather conditions will be further and further into the future. Since 1984, a team of researchers at Colorado State University has been issuing seasonal tropical cyclone forecasts for the North Atlantic Ocean, but these forecasts have had varied accuracy and have been open to significant change (e.g., the 2006 season was forecast to be extremely active but ended up being average).

Long-range forecasting constitutes a lively new frontier for weather scientists and presents some of the toughest research challenges they will have to face. In response to the need and desire for longer-term outlooks, the Canadian Meteorological Centre began producing, through the Meteorological Service of Canada, seasonal temperature and precipitation anomaly forecasts, starting in 1995. The seasonal forecasts are issued four times a year, on the first day of December, March, June, and September. Temperature and precipitation forecasts are made for 1–3 month, 2–4 month, 4–6 month, 7–9 month, and 10–12 month periods. The 1–3 month and 2–4 month forecasts are made using the average of four numerical weather prediction models, while the longer lead-time forecasts are produced with a statistical model (3–12 months). In addition, probabilistic forecasts are produced with the aid of weather models; these indicate the respective probability of the temperature and precipitation being below, near, or above normal. There can be a great disparity between the predicted seasonal forecast and the observed conditions (Figure 2.6 illustrates one such example). The MSC's seasonal webpage (http://weatheroffice.ec.gc.ca/saisons/index_e.html) includes the maps in Figure 2.6 as well as many other forecast maps.

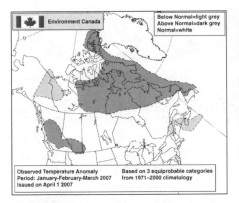 

Figure 2.6 The difference between a seasonal forecast prediction (image on left) and the observed conditions.
Source: Modified from Environment Canada 2007g. © Her Majesty the Queen in Right of Canada, 2007. Reproduced with the permission of the Minister of Public Works and Government Services Canada.

How and What Warnings Are Issued

For the safety of people and property, Environment Canada issues severe weather warnings, watches, and advisories to the public via the media, weather outlets, and Weatheradio Canada (Environment Canada 2006g). A *weather advisory* means that actual or expected weather conditions may cause general inconvenience or concern but are not serious enough to warrant a warning. An advisory may be issued in advance of a warning. A *weather watch* alerts the public to conditions that are favourable for the development of severe weather. A *weather warning* informs the public that severe weather is occurring or that hazardous weather is highly probable (see Table 2.1). Severe thunderstorm or tornado warnings can only be issued less than an hour in advance, while other weather warnings can be issued up to 48 hours in advance (in the case of marine wind warnings, for example).

Canada's severe weather–warning system is far from infallible. For example, in the summer of 2006, the provincial transportation minister for Manitoba urged Ottawa to help fund a new system that would warn Manitobans about severe weather more quickly and easily (CBC 2006g). Prompting this effort were the tornadoes that had killed a woman, injured at least 20 people, and wrecked homes and farms on 5 August 2006. Just a few months prior to this, experts from Environment Canada had criticized the Canadian weather-warning system for being too slow. Today in Manitoba, radio and television stations transmit weather warnings received from the province's Emergency Measures Organization (EMO) within 20 minutes of their receipt. In Alberta, a warning system cuts into all broadcasts within 30 seconds of a warning being issued. In the United States, weather warnings from the National Weather Service are broadcast on radio and television within less than a minute of their receipt (CBC 2006m). Even before the tragic event in Manitoba, Manitobans had been criticizing the storm-warning system (CBC 2005), having experienced damaging thunderstorms that flattened fields, knocked down trees, and tossed around farm equipment.

Canadian provinces are waiting for the federal government to approve a nationwide severe weather alert system, called CanAlert, that would send warnings via

Table 2.1 Examples of weather warnings.

Severe thunderstorm watch	Conditions are favourable for the development of severe thunderstorms with large hail, heavy rain, intense lightning, or damaging winds within the areas and times specified in the watch. You should secure or put away loose objects such as outdoor furniture, put your car in the garage, bring livestock to shelter, and listen for an updated weather report.
Severe thunderstorm warning	A severe thunderstorm has developed, producing one or more of the following conditions: heavy rain, damaging winds, hail of at least 20 mm in diameter, and intense lightning. Severe thunderstorms may also produce tornadoes. The storm's expected motion and developments will be given in the warning. If you are in the area specified, take shelter indoors.
Tornado watch	Conditions are favourable for the development of tornadoes within the areas and times specified in the watch. Be ready to take shelter, preferably in the lower level of a sturdy building.
Tornado warning	One or more tornadoes are occurring in the area specified. The expected motion, development, and duration will be given in the warning. If you are in the path of a tornado, take emergency precautions immediately. If you are near the area specified in the warning, be alert for the development of additional tornadoes or severe thunderstorms.
Freezing rain warning	Expect slippery walking and driving conditions and possible damage to trees and overhead wires due to rain freezing on contact. Avoid travel.
Heavy rain warning	Issued when heavy or prolonged rainfall causes local/widespread flooding. Expect 50 mm of rain over 12 hours or less or 80 mm of rain in less than 24 hours.
Frost warning	Issued when air temperatures are expected to fall to near freezing or below during the growing season, approximately 15 May to 15 October.
Wind warning	Expect winds blowing steadily at 60 km/hr or more, or winds gusting to 90 km/hr or more, for at least one hour. Secure or put away loose objects such as outdoor furniture, put your car in the garage, and bring livestock to shelter.
Marine wind warnings	Small craft warning: Issued if winds of 20–33 knots are forecast Gale warning: Issued if winds of 34–47 knots are forecast Storm warning: Issued if winds of 48–63 knots are forecast Hurricane force wind warning: Issued for winds of 64 knots or more
Dust storm advisory	Issued in prairie provinces when blowing dust caused by high winds has reduced visibility to 1 km or less. Under extreme conditions of widespread zero visibility, this bulletin may be issued as a warning. Dust can impair breathing for people and animals and make travel hazardous.

Table 2.1	Continued
Blizzard warning	Expect snow or blowing snow, with severe wind chill and visibility reduced to less than 1 km, for 4 hours or more. Stock up on heating fuel and food and stay inside until storm has passed.
Heavy snowfall warning	Expect a snowfall of 10 cm or more (15 cm or more in Ontario) in 12 hours or less. Travel may become hazardous.
Winter storm warning	Issued in Ontario when two or more winter conditions reach warning proportions (e.g., wind and snow or freezing rain followed by heavy snowfall). Be prepared to cancel travel plans and stay indoors.
Wind chill warning	Expect combination of very cold temperatures and wind to create outdoor conditions that could be dangerous to human activity. Be prepared to stay indoors.
Cold wave advisory	Temperatures are expected to drop by 20°C or more within 18 hours. Dress warmly and check weather forecast before travelling outdoors.

cellphones, BlackBerrys, pagers, community sirens, teletype machines, and even telephone land lines. They are also awaiting funding for the system (estimated at $400–$500 million) and the completion of a lengthy approval process from the Canadian Radio-television and Telecommunications Commission (CRTC), as CanAlert would involve the use of public radio and television stations. In the meantime, some provinces, such as Alberta, are trying out their own alert systems. In addition, Industry Canada, in partnership with federal government departments, the provinces, and the private sector, is leading the initiative to establish the Canada-wide Public Alerting System (CPAS), which will be based on radio and television broadcasts, cellphones, and the Internet.

In spite of moves for improvement, some criticism has been levied concerning the closure of forecasting centres across the country. Formerly, there were 14 forecasting centres in Canada, but nine were closed or downsized in 2003, leaving Vancouver, Edmonton, Toronto, Montreal, and Halifax to take on extra duties. 'Meteorologists say forecasting won't change much because so much is done by computer, but when weather stations close, weather research in local areas will suffer' (CBC 2003). Gander was reopened in 2006, fulfilling a Conservative government election promise.

Canadians can get a weather forecast from sources other than official government forecasting services. Many Canadians watch their local news stations and have favourite local forecasters. The Weather Network provides continuous coverage of weather events and forecasts 24 hours a day, seven days a week. A smaller percentage of the public regularly check websites such as accuweather.com and wunderground.com, as well as weather blogs such as canadianweather.org. Blogs have recently come to be places where hard-core weather watchers across the country share real-time weather stories in a chat room format.

'Citizen Forecasts'

The reports of weather watchers sometimes constitute the only information available on local severe weather events such as tornadoes or funnel clouds, large hail, flooding, heavy rain or heavy snowfalls, freezing rain or damaging winds (Table 2.2). The Canadian radar network cannot detect the fine detail and localized events that the eyes can see. The volunteer service provides accounts for a large proportion of severe weather reports received by the Meteorological Service of Canada. Since Environment Canada's weather watch program started in 1978, thousands of volunteers have become involved, and as a result of their timely information, many warnings have been issued that might not have been otherwise (Environment Canada 2002a).

Another example of a volunteer weather watcher program is the CANWARN program. CANWARN is a volunteer organization of ham radio operators who report severe weather when they see it to a CANWARN network controller; the controller, using either a special telephone line or the CANWARN webpage, then forwards the reports to Environment Canada's severe weather office in Toronto. At the weather office, the

Table 2.2 What weather watchers report.

Event	Hail	Report
Hail	2 cm dia. and larger; hail damage	Time hail began and ended, depth if covering ground, signs of damage (e.g., flattened crops broken windows, dented cars)
Wind damage	90 km/hr or more	Moderate or severe structural damage, uprooted trees, time wind surge hit, direction and duration
Flooding rain	25–50 mm in 1 hr causing flooding	Roads/fields under water, overflowing creeks/ditches, mudslides, near-zero visibility during cloudburst
Tornado/Funnel	All sightings	Time seen, description of size, distance away, direction of movement if no funnel, whether debris was rising. Was the tornado funnel shrinking or growing? Was it accompanied by sound? Hail? Damage? Was tornado damage discovered in absence of sighted tornado or after the event?
Rotating cloud	All sightings	Time seen, size of lowering (the lowering is a cloud under storm extension [1/2–2 km dia.] on the underside of the rain-free base, usually about 1 km from the edge of the rain core), position relative to main part of the storm
Roaring sound	All occurrences	Time, direction, sky conditions
Unusual sightings, other events (year-round)	All occurrences	Waterspouts, large dust devils, damaging wind, blizzards/snow squalls/fog with near-zero visibility, freezing rain, etc.

severe weather meteorologist combines the data from the satellites and radar with the information from the ground to refine the forecast or prepare a severe weather watch or warning.

Ham or amateur radio operators have long played important roles in their communities, particularly during emergencies. These men and women run very high frequency (VHF) or ultra-high frequency (UHF) radios from their homes, offices, cars, or trucks and thus are in a good position to help when normal lines of communication have been knocked out by a tornado, fire, or explosion. Their roles expanded after the Edmonton tornado of 31 July 1987. The tornado, which had winds of more than 400 km/hr, ploughed through the Alberta city in the mid-afternoon, killing 27 people, injuring 253 others, and causing more than $250 million in damage. The subsequent report on the tornado and the effectiveness of the weather-warning system, known informally as the Hage Report, recommended that Environment Canada solicit the help of amateur radio operators in its severe weather watch and warning program.

Within a week of the report's publication, Environment Canada had trained more than 120 ham radio operators in the Windsor, Ontario, area to detect severe weather. At first, CANWARN operated primarily in southwestern Ontario. Today, there are CANWARN stations in towns and cities from Windsor through Parry Sound on Georgian Bay and points east. There are also CANWARN stations in northwestern Ontario, in places such as Thunder Bay, Fort Frances, Dryden, and Kenora. CANWARN volunteers cover most of the areas of Ontario that are likely to have severe weather. The program is active (to varying degrees) in Manitoba, Saskatchewan, and Alberta (Environment Canada 2005b).

ALTAWATCH is a public service that the Radio Amateur Educational Society (RAES) provides to the Alberta Weather Centre, located in Edmonton. Created by RAES in 1992, it was designed to be the eyes and ears of the Alberta Weather Centre. ALTAWATCH assists in giving early warning to the general public during a severe weather event, providing severe weather reports from volunteers and members of the general public. ALTAWATCH goes into action whenever a severe weather watch is issued by the Alberta Weather Centre. When a warning is issued, volunteers fan out across Edmonton and surrounding communities. They monitor local conditions and report back to Environment Canada, whose monitoring systems are unable to provide local climatic conditions. This service was established after the Edmonton tornado in order to increase alert times during such events. Time can be of the essence during a severe weather event (ALTAWATCH 2006).

Forecasting to the North

In early 2000, concerns were expressed that weather information and warnings were difficult to obtain in many northern communities after regular radio broadcast hours. Weatheradio has subsequently been made available in Cape Dorset, Rankin Inlet, Iqaluit, and Arviat regions of Nunavut since June 2002. Weatheradio is a broadcast service located on the VHF-FM radio band. There are now 168 transmitters located across Canada; they operate on seven different frequencies and provide weather information 24 hours a day, seven days a week. The automated system broadcasts the latest weather observations, provides public and marine forecasts, and issues warnings for specific areas of the country.

One challenge faced by Environment Canada is the variety of dialects found throughout the North. Environment Canada is working with the Nunavut government to see what can be done to find an acceptable Inuktitut weather terminology that would be understood regardless of local dialects. Weatheradio now broadcasts in Inuktitut, and people can also call a 1-800 number for their forecast in Inuktitut. Weatheradio is available in every province and territory.

Chapter Summary

Canadians are probably more aware of weather conditions today than they ever were before. Severe weather can be a challenge for forecasters given the low statistical frequency of such events, the wide variation in model predictions, and the public's heightened awareness of and attention to the details of the forecast. Weather forecasting in general has changed significantly since the early part of the twentieth century. Today, forecasting involves the use of radar, satellite technology, and complex weather models. Long-range forecasting constitutes a lively new frontier for weather scientists and presents some of the toughest research challenges they will have to face. Environment Canada issues severe weather warnings, watches, and advisories to the public.

Chapter Three

Severe 'Winter' Weather

'I knew I was in trouble when I saw my neighbour walk past my second-story window.'

Resident of Iqaluit, NT, in reference to the huge
snowdrifts caused by the 14–15 February 1988 blizzard

Objectives

- To provide an overview of the variety of severe winter weather conditions that exist across Canada
- To introduce processes associated with severe winter weather in Canada and the storm tracks they tend to follow
- To provide an overview of severe wind systems (including Wreckhouse winds, Alberta clippers, chinooks, and nor'easters)
- To describe the spatial trends and characteristics of snow accumulations, blizzards, and ice storms across Canada
- To explain the characteristics of cold weather and wind chill
- To discuss the implications of climate change and severe winter weather conditions in Canada

Introduction

Winter weather more than any other weather condition is characteristically Canadian, perhaps as typically Canadian as hockey and the maple leaf. One thing that this chapter will underscore is the fact that severe winter storms are far from rare in Canada. Our history is littered with examples. Moreover, 'winter' weather isn't restricted to the calendar dates of the official season; we can have snowstorms in the fall and well into the spring in some parts of the country. In fact, the median start and end dates of snow cover across Canada are well outside of the limits of the winter season (see maps in the colour insert section). The median start date for continuous snow cover across Canada is usually prior to December and for more northerly latitudes can be as early as August and September. The median end date for continuous snow cover illustrates more regional variation, with some areas typically losing snow cover before spring officially begins (e.g., southern Alberta) and others having snow cover well past the official winter season (e.g., northern Ontario, northern Quebec, and higher latitudes).

The fall of 2006 and spring of 2007 presented good examples of severe winter weather conditions occurring well outside of the winter season calendar dates. In October, a 'historic' snowstorm hit the Niagara, Ontario, region, virtually shutting down the entire town of Fort Erie (CBC 2006d). Dumping 30 cm, this snowstorm was part of a large system that brought record amounts of snowfall and closed down the Peace Bridge, which connects Canada and the United States. Since records first began to be kept in the 1870s, the area had never seen as much snow fall over a period of a day and half as early as October. Meanwhile, trick-or-treaters in southern Manitoba had to bundle up to go out in 15 cm of snow. On the west coast, heavy snow (30–60 cm) in northern British Columbia caused power lines and trees to come down, leaving 15,000 people without electricity (CBC 2006i). Barely a month later, on 25 November 2006, a rare storm blanketed the province's southern coast with 30 cm and was blamed for at least one death as well as power outages affecting thousands of people in the Lower Mainland (CBC 2006j). On the east coast, rush hour commuters in Halifax were taking two to three hours longer than normal to get home on Monday, 4 December 2006 (Figure 3.1), as the season's first snowstorm (a nor'easter) paralyzed the city under 19 cm of snow, taking many people by surprise (*Chronicle-Herald*, 2006). On the other end of the seasonal spectrum, children were digging their chocolate Easter eggs out of the snow in many parts of Nova Scotia on 8 April 2007. Almost 31 cm of snow fell in Halifax, surpassing the entire total accumulation for the month of January (28.5 cm) (*Chronicle-Herald* 2007c).

It is not uncommon for many parts of the country to be under winter weather warnings or watches for much of the winter season. Every year, Environment Canada issues 14,000 warnings in addition to 1,100,000 forecasts (Environment Canada 2001). On 13 February 2008, Environment Canada issued weather warnings for all parts of Canada except British Columbia, Yukon, and southern Ontario. Snowstorms were forecast for Quebec and most of the Maritime provinces. New Brunswick was expecting freezing rain, the Prairies were experiencing hazardously low temperatures, and parts of southeastern Alberta and Saskatchewan were anticipating blizzard conditions. Wind chill warnings were issued for Manitoba, with temperatures as low as −50°C expected, and blizzard and wind chill warnings were issued for parts of the Northwest Territories and Nunavut. In the midst of it all, many Canadians would have found the Australian government's warnings about travel to Canada rather amusing. According to the official website of the Government of Australia (2008), Australians were urged to 'exercise caution' when travelling to Canada in part because of 'heavy snowfalls and ice in the winter' and 'the wind chill factor [which could create] dangerously cold outdoor conditions.'

Canadian provinces and municipalities expend considerable resources on measures to prevent unsafe road conditions associated with severe winter weather from developing. The province of Nova Scotia spent over $43 million on snow removal and ice control during the winter of 2006–7 alone. The need for such expenditures is obvious. More than 100 fatalities each year can be attributed to severe winter weather conditions, such as bitter cold and winter storms. This is more than the number of Canadians killed by tornadoes, thunderstorms, lightning, floods, and hurricanes combined (Environment Canada 2007k).

Figure 3.1 Spring Garden Road, Halifax: First winter storm of the season, 4 December 2006.
Source: Cathy Conrad.

Causes of Storms

The most common cause of severe winter weather in Canada is the extratropical, or mid-latitude, cyclone (Stewart et al. 1995; Brun et al. 1997). Such cyclones can produce the following:

- Heavy snowfall and blowing snow (e.g., blizzards)
- Freezing rain
- Severe cold snaps
- Severe Atlantic and Pacific coastal storms

A study of the entire Northern Hemisphere indicates that approximately 234 significant extratropical cyclones form each winter (Gulev et al. 2001).

Extratropical cyclones are low-pressure weather systems that occur in the middle latitudes of the Earth (typically between 30° and 60° latitude) and that have neither tropical nor polar characteristics. Although their name might conjure up images of heavy gales and precipitation, these systems are not necessarily associated with those conditions. Extratropical cyclones and their counterpart high-pressure systems (anticyclones) drive the weather over North America and can produce conditions ranging from mild showers and overcast skies through to intense storms. Several such storms are typically present over Canada at any given time during the winter period (Figure 3.2).

Mid-latitude cyclones develop when a divergence of air at elevations above the Earth's surface removes more air than can be brought in by a convergence of air at the surface and thus surface air pressure starts to drop. This creates a horizontal pressure gradient, and cyclonic circulation begins—the initiation of a storm system (Moran et al. 1997). The westerly flow of air in the upper atmosphere then steers and supports the cyclone as it progresses through its life cycle. In fully developed extratropical cyclones, heavy precipitation thunderstorms and thundersnows often form to the northwest of the low-pressure system, in a region known as the *comma head*. Atmospheric pressure can drop rapidly as air continues to rise from the surface. Between 4 and 5 January 1989, an extratropical cyclone south of Atlantic Canada brought surface pressures down to 928 mb (millibars), roughly equivalent to levels in a category 4 hurricane (Masters 2006).

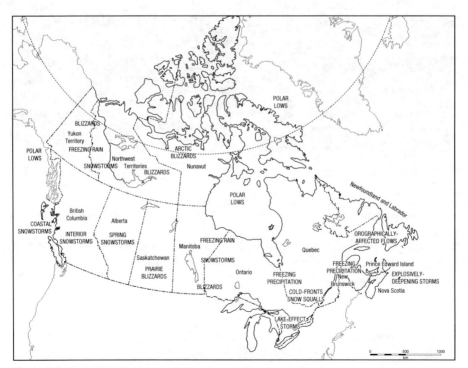

Figure 3.2 Types of winter storms and their most common locations in Canada.
Source: Modified from Stewart et al. 1995.

Extratropical cyclones usually affect large areas (1000s of km^2) and occur relatively frequently. They can be responsible for considerable property damage and many deaths. For example, in March 1993, a mid-latitude cyclone migrated up the east coast of the United States and Canada, producing a severe blizzard and killing more than 240 people—three times as many as the number killed by Hurricanes Hugo and Andrew combined. The highest winds ever recorded in Atlantic Canada (233 km/hr) occurred in this storm (Bowyer 1995). At one point during the storm, over 3 million people were left without electricity because of high winds and falling trees (Smith 1996; Brun et al. 1997), a reminder of society's vulnerability to the different features of severe weather systems.

Some winter storms in Canada are topographically affected. The passage of an extratropical cyclone through topographically varying terrain can lead to the development of strong surface winds. Atlantic Canada and the west coast are particularly prone to this effect. Mountain ranges along the west coasts of Cape Breton Island and Newfoundland are aligned southwest to northeast, and when cyclones move relatively close to the coast, a strong southeasterly gradient wind is set up perpendicular to these mountain ranges. The southeasterly wind creates a mountain wave effect, which often brings severe downslope windstorms to the lee of the mountains (90–200 km/hr). These winds occur several times a year and are well known in local lore. In the Wreckhouse region of Newfoundland, they are called 'Wreckhouse winds' (see section on Wreckhouse winds below for more details), while the Acadians in Cape Breton refer to them as 'les suêtes'.

Additional causes of severe winter weather include cold-front squall lines or snow squall lines, which are sometimes associated with an intense, rapidly moving cold front. These snow squalls produce conditions of sudden near-zero visibility in snow and blowing snow that may last for only a few minutes. The squalls tend to occur during the middle of winter (January and February). Southern Quebec (two to four occurrences per year) and the Atlantic provinces (up to 10 occurrences per year) are particularly prone to these features (Stewart et al. 1995).

Lake-effect storms are another important origin of winter systems, especially in areas near the Great Lakes, the Gulf of St Lawrence, regions of the Arctic, and areas southeast of the Manitoba lakes. Lake-effect snow is produced by the passage of cold air over a relatively warm body of water. The potential for the formation of lake-effect snowstorms is primarily determined by the temperature difference between the water surface and the overlying cold air mass. These systems often result in narrow, intense bands of heavy snow (snow squalls) (Stewart et al. 1995).

Winter Storm Tracks

Winter storm systems in Canada tend to be more intense and more frequent than summer storm systems because there are greater temperature differences between northern and southern latitudes during the winter months. Winter low-pressure storm systems favour certain tracks, along which broad areas of cloud cover, snowfall, rain, high winds, and sometimes freezing precipitation will be found. The tracks that have an impact on Canada are generally named according to their points of origin (Figure 3.3).

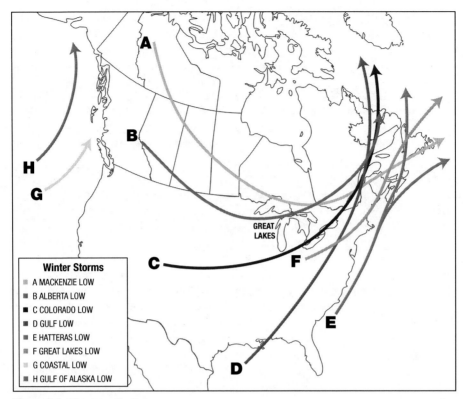

Figure 3.3 Winter storm tracks.
Source: Modified from Nav Canada.

Mackenzie Low storms develop in the Mackenzie River Valley of the Northwest Territories and follow a southeastward track, tending to the north of the Prairie provinces. Occasionally they will impact more southerly areas of Saskatchewan and Manitoba during the winter. Storms associated with the Alberta Low form on the lee of the Rocky Mountains in Alberta and move across the Canadian Prairies, northern Ontario, and northern Quebec. Although they tend not to generate a lot of precipitation, they can be enhanced by an inflow of warm moist air moving up the Atlantic seaboard. When such conditions exist, these storms can intensify rapidly and bring heavy winter snowfall to southern Ontario and Quebec.

Colorado Lows also develop to the lee of the Rocky Mountains but in this case over southern Colorado. They generally track northeastward over the Great Lakes, where they can intensify and bring heavy snowfalls and strong winds to Ontario and Quebec. Cold air moving across the relative warmth of the vast moisture supply of the Great Lakes can serve as yet another source for winter low-pressure systems. Such systems tend to track either to the northeast across eastern Ontario, Quebec, and Labrador or move towards the east and the Atlantic, where they can re-intensify off the east coast of the United States.

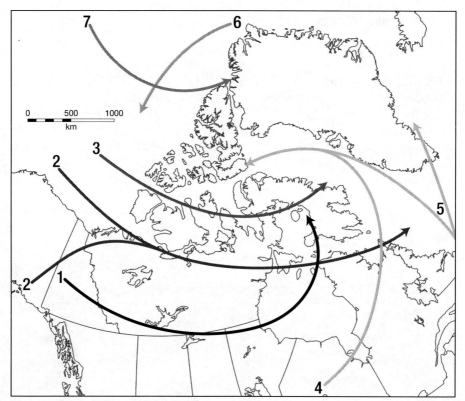

Figure 3.4 Winter storm tracks that impact Nunavut and the Arctic region.
Source: Modified from Nav Canada.

Hatteras Lows originate off the coast of Cape Hatteras in North Carolina. When cold arctic air from the north meets the warmth of the Gulf Stream, a strong temperature gradient develops. The result is often the rapid development of a winter weather system. These systems tend either to track across Nova Scotia and Newfoundland or to move across Quebec and towards Ungava Bay. When these systems rapidly intensify (e.g., when air pressure drops 24 hPa [hectopascals] in 24 hours or less), they are often referred to as 'bombs', the consequence of which are severe winter weather conditions across eastern Canada. These storm systems are often difficult to forecast.

The systems that affect British Columbia develop over the Pacific Ocean and move towards the coast as either Gulf of Alaska Lows or Coastal Lows. The former tend to remain offshore, but they sometimes drift back towards the southeast and regain strength towards the coast of British Columbia. The systems that affect Nunavut and the Arctic are even more complex (Figure 3.4). Lows originating over the Prairies (Track 1), the Pacific and Arctic (Tracks 2 and 3), southern Ontario and Quebec (Track 4), the Maritimes (Track 5), the North Atlantic (Track 6), and across the Arctic Basin (Track 7) all have the potential to bring strong winds and blizzard conditions to Nunavut and the Arctic.

Wreckhouse Winds

In 2004, the *Canadian Oxford Dictionary* added a new expression that describes a wind-related extreme weather phenomenon. That expression was *Wreckhouse winds*, defined as 'extremely strong winds which blow across Cape Ray from the Long Range Mountains in Newfoundland' (Figure 3.5). Wreckhouse winds can actually reach hurricane speeds. During the Newfoundland winter, motorists who set out on the Trans-Canada Highway in this region often face the fury of the Wreckhouse winds and have to decide whether to delay their journey or push through and hope for the best. Environment Canada and other forecast providers issue special wind forecasts and warnings for the Wreckhouse region.

Stories about the damage caused by the Wreckhouse winds go back more than a hundred years. In January 1900, a powerful storm struck the southwest region of

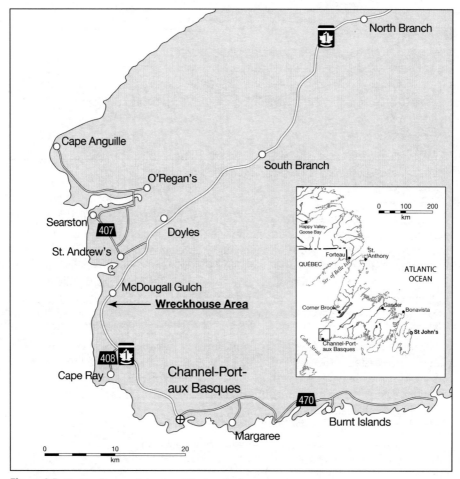

Figure 3.5 The Wreckhouse wind region of Newfoundland.

Source: Modified from Keith C. Heidorn, The Weather Doctor, http://www.island.net.com/~see/weather/doctor.htm

Newfoundland, blowing a train right off the tracks. During the early 1930s, the managers of the Newfoundland Railway became aware of a local farmer and trapper named Lauchlan ('Lauchie', pronounced 'Lockie') MacDougall. Lauchie, born in 1896, was a long-time resident of the Wreckhouse region. According to reports, Lauchie could sense the signs of an approaching storm and the intense winds that accompanied it. The railway managers decided it was worth a shot to hire him as an observer—a 'human wind gauge'—at $20 a month. It would be his job to warn the nearby Port-aux-Basques rail office of the onset of extreme winds. This would allow the railway managers to halt trains before they reached the dangerous Codroy Valley section of the track. Reportedly, on one occasion a train conductor ignored Lauchie's 'forecast' and continued through the valley. When 22 cars of that train were blown off the tracks by the Wreckhouse winds, Lauchie's stature rose and he became a local legend. He would provide warnings until his death in 1965. Lauchie MacDougall is credited with likely saving many lives and much property over the years by delaying hundreds of trains because of treacherous wind conditions. Today, his empty home still stands along the Trans-Canada Highway, an Environment Canada remote anemometer rising beside it to monitor the Wreckhouse winds (Heidorn 2005a).

Strong winds are common in southwest Newfoundland at any time of the year, although they become more frequent and of higher velocity during the winter and spring when strong Atlantic storms slam into the island. When intense storm systems approach Newfoundland from the south (often as coastal nor'easters), the circulation around the low-pressure centre commonly pushes strong southeasterly winds ahead of it. The Wreckhouse area is a stretch of flat, barren land that extends along the edge of the Long Range Mountains, the highest range on Newfoundland. As southeasterly winds blow across the region, the air is funnelled down the long valleys that are cut into the sides of the mountains. When the winds reach the end of these valleys, the air is no longer confined to the valley walls and quickly extends across the landscape. Gale- to hurricane-force winds sweep downslope from the ridge summits through the narrow valleys and stunted forest at speeds that can gust in excess of 160 km/hr, the equivalent of a category 2 hurricane. Although there have been indications that the frequency of Wreckhouse wind events have increased in the past several years, no long-term trends are evident (Catto et al. 2006).

Alberta Clippers

A well-known wind phenomenon that originates in Alberta but whose impact extends well beyond that province is the so-called Alberta clipper. An *Alberta clipper* (sometimes referred to as a Canadian clipper in the United States) is defined by the American Meteorological Society (AMS) as 'a low pressure system that is often fast-moving, has low moisture content, and originates in western Canada (in or near the province of Alberta). In the wintertime, it may be associated with a narrow but significant band of snowfall, and typically affects portions of the plains states, Midwest, and East Coast' (AMS 2008). These systems get their name from the Canadian province of Alberta as well as from the clipper ships of the nineteenth century, some of the fastest-moving ships of their time. Alberta clippers generally form east of the Rocky

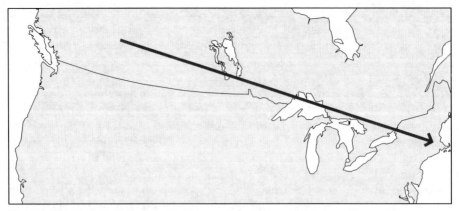
Figure 3.6 The average trajectory of an Alberta clipper.

Mountains and then travel towards the southeast under the push of the northwesterly jet stream into parts of the US northern plains and the central Canadian provinces (Figure 3.6). Because they tend to be fast-moving systems and their track takes them far from significant moisture sources (e.g., the Atlantic and Pacific Oceans), the average Alberta clipper rarely brings a significant amount of precipitation. They can still be fierce storms, however, with wind speeds up to 65 km/hr and gusts up to 100 km/hr. Even with small amounts of snow already on the ground, these strong winds can generate whiteout and blizzard conditions (Heidorn 2005a). Although most regions do not receive much in the way of new precipitation from these systems, the southern and eastern shores of the Great Lakes receive enhanced snowfall from clippers during the winter because of the lake-effect snow, which can add substantially to the overall snowfall. Most clippers occur between December and February, but they have been known to occur in November.

Alberta clippers are often associated with extremely cold temperature outbreaks, with drops in temperature of 16°C in as little as 10 to 12 hours not being uncommon. With winds in advance and during a clipper as high as 56–72 km/hr, wind chill values drop into the −30°C to −45°C range (Wikipedia 2007a). Occasionally, an Alberta clipper crosses the Appalachians heading southward and becomes re-energized by warm Atlantic coastal waters to be reborn a nor'easter (Heidorn 2005a).

Nor'easters

A *nor'easter* is an intense extratropical cyclone that tracks up the east coast of North America. Nor'easters aren't as powerful as hurricanes, but they can be very destructive and can occur at any time of year, although they are most common between October and April. These systems tend only to strike land along the North American Atlantic coast, from Cape Hatteras north to the Canadian Maritimes and Newfoundland, bringing gale-force winds (63–87 km/hr), heavy precipitation, and heavy surf along the coast. Nor'easters move towards the northeast and, if centred offshore, can produce strong onshore winds that can cause considerable coastal erosion, flooding, and property damage. They have become infamous along the Atlantic coast of Canada, notable examples being the Blizzard of '88, the 'Perfect Storm' of 1991, and the snowstorm of February 2004, dubbed 'White Juan' (Heidorn 2005a).

Most nor'easters develop from systems that have originated very far away, such as from Alberta clippers that move to positions off the Atlantic coast. When these systems encounter the warmer waters of the Atlantic Ocean, they regenerate, many moving north along the coastline. A nor'easter can be three times the diameter of an average hurricane, and because it often moves parallel to the coast, its onshore winds may impact more than 1500 km of coastline. A less intense but slow-moving nor'easter can actually be more disruptive than a more intense but fast-moving one. Winds from the slow-moving system blow in the same onshore direction for a longer period than the continually shifting winds associated with the fast-moving system (Moran et al. 1997).

Chinook Winds

While in and of themselves chinook winds are not related to outbreaks of severe weather, they can contribute to severe weather events. The *chinook* is a warm and dry downslope breeze that develops when relatively mild air above the surface becomes compressed as it descends on the leeward slopes of mountain ranges. It is drawn downslope by strong west winds associated with cyclones and anticyclones well east of the mountains (Moran et al. 1997). For every 1000 m of descent, the air temperature rises about 10°C, and therefore the air that flows down the slopes of the Rockies undergoes considerable warming. The chinook's effect can be dramatic. In Pincher Creek, Alberta, for example, a chinook caused the temperature to rise 21°C in four minutes on 6 January 1966.

'Chinook' is a Blackfoot First Nation word that according to tradition means 'snow eater'. The term is appropriate, since the warm dry winds assist in melting snow cover. On 25 February 1986, a chinook descended on Lethbridge, Alberta, with winds gusting to 166 km/hr, fully removing a snowpack 107 cm in depth within eight hours. Lethbridge was left with substantial wind damage and new lakes standing in the surrounding fields and pastures (Heidorn 2005a). Calgary, too, is well known for getting many chinooks. The Bow Valley in the Canadian Rockies west of the city acts as a natural wind tunnel as it funnels the chinook winds. In February 1992, Claresholm, Alberta, hit 24°C, one of Canada's highest February temperatures (Wikipedia 2007b). In Canada, chinooks are most prevalent in southern Alberta, especially in a belt from Pincher Creek and Crowsnest Pass through Lethbridge, an area that has an average of 30 to 35 chinook days per year. They become less frequent farther south in the United States and are not as common north of Red Deer, but they do occur as far north as Grande Prairie in northwestern Alberta and Fort St John in northeastern British Columbia, and as far south as Albuquerque, New Mexico (Figure 3.7).

Chinooks occur year-round, although their warming effect is more apparent in colder weather. They can last less than an hour or for several days. Chinooks can be

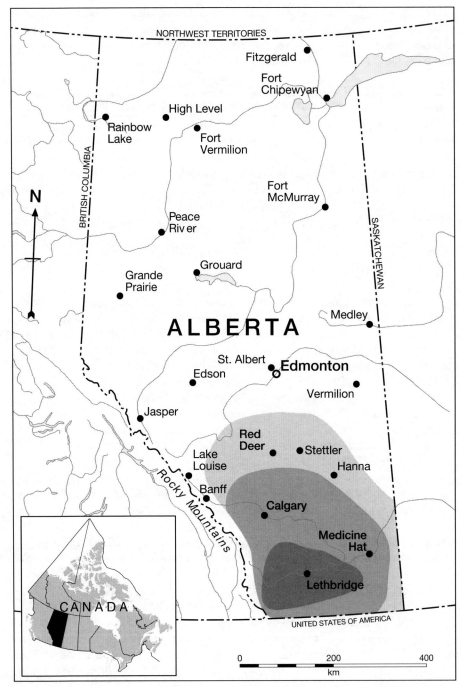

Figure 3.7 Region of most intense chinook winds.
Source: Cathy Conrad.

welcome during the winter, giving a respite to residents of the cold Prairies, but in other seasons, the searing dry winds can cause vegetation to deteriorate, raise soil into dust storms, and rapidly increase the danger of grass and forest fires. For many living under the chinook influence, its winds bring debilitating physical effects, ranging from sleeplessness to anxiety and severe migraine headaches. Recent studies also suggest that chinook winds rolling off the mountains can damage aircraft at cruising altitudes by generating turbulence powerful enough to rip an engine from a jet. Chinook winds can be a fire hazard in the Canadian Prairies during dry periods (Heidorn 2005a).

One of the most striking features of the chinook is the *chinook arch*. This is a band of stationary stratus clouds caused by air rippling over the mountains due to orographic lifting. To those unfamiliar with the chinook, the chinook arch may sometimes look like a threatening storm cloud, but chinooks rarely produce rain or snow. It addition to the arch, chinooks also create stunning sunrises and sunsets (Wikipedia 2007b). (See photo 'A chinook arch', in colour insert section.)

Related to the chinook is the *squamish,* a strong and often violent wind that occurs in the fjords, inlets, and valleys of British Columbia. Squamishes move in a northeast-southwest or east-west direction, causing cold polar air to funnel westward, the opposite of its usual direction on the coast. In the winter, these winds can create wind chills of –20°C to –30°C, high by coastal standards (Wikipedia 2007d). Squamishes lose their strength when they move beyond the confining fjords and are not noticeable more than 25 km offshore.

On the Lower Mainland and eastern Vancouver Island, where they are sometimes referred to as 'outflow winds', squamishes are especially noticeable in the winter when masses of cold arctic air flow down to the sea through the canyons and lower passes and cross the Strait of Georgia (Wikipedia 2007d). The town of Squamish, BC, is named for the wind. Squamish means 'Mother of the Wind' in Coast Salish, which is testimony to the winds that rise from the north before noon and blow steadily until dusk, making Squamish a top wind-surfing destination.

Snow

Most Canadians are used to snowstorms and have found ways to adapt to their adverse conditions (Figure 3.8). In central Newfoundland, on 9 March 1988, a man was spotted standing on a ladder trying to shovel out his buried car after a blizzard had deposited 38 cm of snow. In Cornwall, Ontario, during a blizzard on 4 March 1971, citizens who made it to work received a certificate dubbing them 'Champions of the Big Blow'. The storm left 53 cm of snow in 30 hours, resulting in as many as 500 tractor-trailers marooned on Highway 401 with snow up to the tops of their cabs (Environment Canada 2004j). During the 18–19 February 2004 snowstorm in Halifax, a pregnant woman had to be lifted over a snowbank in a forklift so that she could get to the hospital in time.

Snow is a hallmark of the Canadian winter. Children talk about 'snow days' and outdoor enthusiasts await snowfall for outdoor activities, while on the other end of the spectrum, 'snow birds' leave the country to escape the snow. Snow isn't usually

Figure 3.8 The aftermath of a large snowstorm in Newfoundland. Snow accumulates in deep drifts near large objects like this house.

Source: The Canadian Press/Tony Brown.

considered a problem on the west coast, where Vancouver and Victoria each have an average annual snowfall amount of less than 60 cm. In 1996, however, an atypical storm dumped a record one-day snowfall of 41 cm on Vancouver and 64 cm on Victoria. Transportation ground to a near standstill. Ambulances, fire trucks, and other emergency services had trouble reaching people in need. Radio stations sent out urgent requests to the cities' few snowmobile owners, asking for their assistance in getting people to emergency services. Canadian Forces soldiers were sent in to clear roads and rescue hundreds of stranded motorists (Environment Canada 2004i).

Snow accumulations have a variety of socio-economic effects that can be both positive and negative. Although many Canadians bemoan a forecast calling for more snow, others welcome its accumulation. In the Prairie provinces, snow is an important component of annual runoff, water supplies, and water management. On the other hand, rapid melt of snow cover can be a major cause of floods in many parts of the country.

Bailey et al. (1997) have described the spatial variation in snowfall amounts in Canada. Central and northern regions tend to receive the least snowfall, and western and eastern regions receive three to four times as much as Saskatchewan and Manitoba. The low snowfall amounts in north and central parts of the country can in part be explained by the blocking effect of the western Cordillera mountains on the air masses coming from the Pacific, the lack of frontal systems in these regions, and the distance of the regions from moisture sources (Bailey et al. 1997). The largest snowfall totals are on the northwestern Pacific coast, high altitudes in British

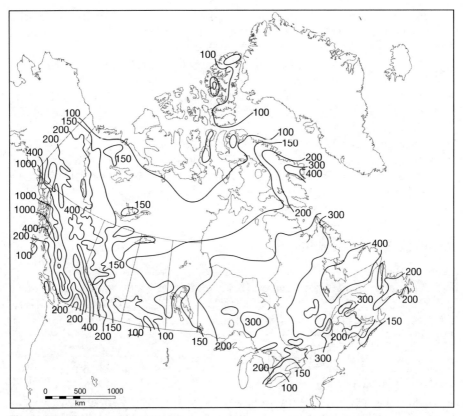

Figure 3.9 Average annual snowfall (cm).

Columbia, Ontario northeast of Lake Superior, and the higher elevations of Quebec and Labrador and Newfoundland (Figure 3.9). These larger totals are likely due either to orographic cooling or to the proximity of these areas to moisture sources. Spatial variability in snowfall amounts can be accounted for by shifting storm tracks (see previous section) as well as by rain-snow variations. Parts of New Brunswick often experience heavy snowfall, while Nova Scotia (particularly along the Atlantic coast) receives precipitation in the form of rain. This can be attributed to the cyclonic flow around a low-pressure system: a relatively warm flow moves up from the east and south of the system, impacting parts of coastal Nova Scotia, while the colder flow of air coming from the north and west brings snow to parts of New Brunswick.

In the winter months, much of the Canadian Arctic can be viewed as a northern desert. Very little moisture is imported into the region, and for that reason, coupled with the fact that cold arctic air cannot hold much moisture, the snowfall in the region is relatively light compared to that in more southerly areas of the country (Stewart et al. 1995).

While certain regions across the country have the largest accumulated amounts of snowfall (the greatest depths), individual locations in Canada, as noted above, can have large single snowfall events. Not only has snowfall varied spatially, but there has also been considerable temporal variation over the past century (Groisman and Eastlering 1993; Groisman et al. 2004). Annual precipitation increased in southern Canada (south of 55°N) by 13 per cent from 1891 to 1990, and there was an increase of up to 20 per cent in annual snowfall and rainfall during 1960–90 in latitudes 45°N –55°N. The precipitation increase has also been marked by a redistribution from snow to more rainfall. The main area for the increase is eastern Canada. Moreover, Groisman and Easterling (1993) report that snowfall between latitudes 45°N and 55°N is closely related to temperature changes, and they can 'foresee a future decrease in snowfall for this zone' (201). Groisman et al. (2004) find that there has been a decreasing trend in spring season snowfall and a general shortening of the snow season and that this is likely associated with spring season warming. Many studies in the United States indicate that there has been a trend towards more precipitation falling as rain than as snow (e.g., Knowles et al. 2006).

Notable Canadian Snowstorms

With the experience of Hurricane Juan still fresh in people's minds, Nova Scotians endured another exceptional storm on 18 and 19 February 2004—a storm that many dubbed 'White Juan'. On 18 February, a low-pressure system developed off the eastern seaboard of the United States and intensified rapidly as it tracked northeastward. By the morning of 19 February, with the storm centre located south of the Maritimes, heavy snow and strong winds had spread over all areas of Nova Scotia, Prince Edward Island, and southeast New Brunswick. A province-wide state of emergency was declared in Nova Scotia as the massive snowfall paralyzed the region. The Halifax Regional Municipality (HRM) imposed traffic curfews, and businesses and schools were forced to close, as many roads could not be cleared for days. By 20 February, the storm had left behind a large swath of snowfall accumulations in the range of 50–70 cm, with a few pockets of 80 cm or more. Canadian Forces Base Shearwater recorded a whopping 95 cm of snow during this event. With nowhere to put the excessive amounts of snow, HRM obtained special permission from the Department of Fisheries and Oceans to dump snow in Halifax Harbour. Winds during the height of the storm were generally in the 60–80 km/hr range, with gusts near 120 km/hr in exposed areas. The wind and heavy snow combined to produce near-zero visibilities and huge snowdrifts (Environment Canada 2004j).

One of Montreal's worst snowstorms occurred on 4 March 1971, when 43.3 cm of snow fell on the city. This storm demonstrated the vulnerability of major cities to severe climate events. A total of 382.7 cm of snow, the heaviest snowfall ever recorded for the city, blanketed Montreal during the winter of 1970–71 (CRIACC 2001). There had been other major snow events in Montreal prior to 1971. What was so special about the storm of March 1971 and why does it warrant the title 'Snowstorm of the Century'? The most notable characteristic of this storm wasn't the amount of snow that fell, but rather the combination of several factors, including the snowfall intensity, the wind speed, and the depth of snow already on the ground (56 cm) when the storm began.

Table 3.1 provides information on some of the worst snow events to strike in Canada from 1916 through to 2004.

Table 3.1 Notable snowstorms in recent Canadian history.
Source: Modified from Environment Canada 2004i, 2005f.

Location	Date	Conditions
Victoria, BC	1–3 Feb. 1916	A 38-hour snowstorm shuts down the city under 78.3 cm of snow.
Listowel, ON	28 Feb. 1959	An accumulation of heavy snow causes the roof of an arena to collapse during a hockey game. Several inside the arena are killed.
Southern Alberta	April 1967	A series of intense snowstorms drops a record 175 cm of snow on southern Alberta. The deep snow blocks roads, closes schools, and cuts off power. Army units are dispatched to assist in the snow clearing, while food, fuel, and cattle feed are airlifted into the province. Thousands of cattle, unable to forage for food in the deep snow, perish on the open range.
Montreal, QC	4 March 1971	Montreal's 'Snowstorm of the Century' dumps 47 cm of snow and causes 17 deaths. Electricity is cut for up to 10 days in some areas. The city hauls away a total of 500,000 truckloads of snow.
Off the coast of Newfoundland	14–15 Feb. 1982	The ocean-drilling rig Ocean Ranger overturns and sinks during an intense snowstorm with winds in excess of 100 km/hr off the east coast of Newfoundland. Most of the lifeboats crash into the pillars of the rig and sink or are flipped over by the waves. There are no survival suits. When the supply ship reaches the scene, less than an hour after being dispatched, only one lifeboat is found afloat with living people on it; the lifeboat slowly rolls over, eventually killing all aboard. A total of 84 die. The severe storm also causes the Soviet freighter *Mechanic Tarasov* to sink, with a loss of 33 lives.
St John's, NL	10 Feb. 1987	A few days after the year's biggest snowfall dumps 30 cm on St John's, a section of a mall roof collapses under the weight of the snow. Seven people are injured but none seriously.
Moncton, NB	1 Feb. 1992	A major winter storm blasts the Maritimes, dumping 60–90 cm of snow across New Brunswick and Nova Scotia. Moncton registers 160 cm of snow, setting its all-time single-storm record.
Victoria, BC	28–29 Dec. 1996	80 cm of snow fall in 24 hours and 125 cm in five days, with cleanup costs exceeding $200 million (including a record insurance payout for BC of $80 million).
Toronto, ON	2–15 Jan. 1999	Toronto's 'Snowstorm of the Century' dumps nearly a year's amount of snow in less than two weeks. In all, the city records the greatest January snowfall total ever, 118.4 cm, and the greatest snow on the ground at any one time, 65 cm. The storms cost the city nearly twice the annual budget in snow removal.
Tahtsa Lake, BC.	11 Feb. 1999	Canada's snowiest day (145 cm) is recorded at Tahtsa Lake, located in the Whitesail Range of the Coast Mountains. This record is still well below the world record (192 cm), set at Silver Lake, Colorado, in 1921.
Nova Scotia and PEI	19–20 Feb. 2004	'White Juan.' See full description in text.

Blizzards

There doesn't need to be a lot of snow on the ground for drifting and blowing snow conditions to occur. So-called whiteout conditions contributed to a 50-vehicle pileup north of Toronto on 20 January 2008, although there was only a minimal amount of snow on the ground (CBC 2008). As wind speeds increase, snow can be entrained by the wind and lifted above the land surface, progressing from drifting conditions (snow particles raised to a height of less than 2 m) to blowing snow (restricted horizontal visibility and snow particles raised more than 2 m). Blizzards can be differentiated from snowstorms or blowing snow in that blizzards have wind speeds of 40 km/hr^{-1} or more, with snow or drifting snow and wind chill factors of up to $-25°C$, and last for at least four consecutive hours (Environment Canada 2001). According to Environment Canada, blizzards pose the greatest threat of any Canadian weather phenomenon. They are most common in northern Yukon, the southern Prairies, Atlantic Canada, and Nunavut, although no region of Canada is completely immune (Environment Canada 2001). The feature that makes blizzards so hazardous is limited visibility (less than 1 km in blowing or drifting snow). There doesn't even have to be a snowfall for there to be a blizzard. In fact, some of our worst blizzards have occurred with very little new snow—just old snow being pushed around. At least one or two people and much livestock perish from exposure to blizzards each year (Stewart et al. 1995). Table 3.2 provides a list of some of the most severe Canadian blizzards between 1941 and 2006.

To achieve an understanding of the frequency of these events, we must view blizzards in terms of the number of days in a year that an area experiences blowing snow and the annual depth of snowfall. Figure 3.10 shows the average annual number of days with blowing snow across Canada. The Arctic has the most days with blowing snow, with areas above the treeline providing an ideal environment for wind to reach the sustained speeds required to maintain snow in motion. The maximum in western Canada occurs in southern Saskatchewan and Manitoba. The eastern shore of Lake Huron receives the highest incidence in eastern Canada, mainly because of lake-effect snow from the Great Lakes.

Blizzards fortunately tend to be short-lived, and they display a variation in occurrence over the winter. About 61 per cent of blizzards last less than 10 hours, and about 90 per cent end within 18 hours. Only 5 per cent last more than 24 hours. January has the most blizzards, but no storm in this month has ever lasted more than 30 hours (Stewart et al. 1995).

Blizzards are typically classified by the source region of a low-pressure system (Lawson 1987). The most frequent blizzard-producing storm is the Alberta clipper (discussed above). This type of storm produces approximately five blizzards over the Prairies each winter and accounts for about 65 per cent of all blizzards in Canada. Alberta clipper storms affect all of the Prairies, but they have their greatest impact on Saskatchewan. Blizzard conditions associated with these systems usually last less than 12 hours. Low-pressure systems initiated over Colorado (Colorado Lows) produce about 15 per cent of the blizzards experienced over the Prairies. The effects of these storms are usually restricted to southern Manitoba and northwestern Ontario. Snowfalls can be fairly heavy, and the blizzard conditions behind these lows can last for one to three days (Heidorn 2005a). According to Lawson (2003b), there is some

Table 3.2   Notable blizzards in recent Canadian history.

Sources: Modified and compiled from Burrows and Treidl 1979; Phillips 1990; Stewart et al. 1995; Brun et al. 1997; Environment Canada 2004l, 2005f; PSC 2006a.

Location	Date	Conditions
Prairies	14 March 1941	Blizzard lasts only 7 hours, but winds are in excess of 100 km/hr. The storm results in 76 deaths.
Toronto	11 Dec. 1944	Two days; 57 cm of snow
Central Canada	30 Jan. 1947	A 10-day blizzard buries towns and trains from Winnipeg to Calgary, causing some Saskatchewan roads and rail lines to remain plugged with snow until spring. Children step over power lines to get to school and build tunnels to get to the outhouse. A Moose Jaw farmer has to cut a hole in the roof of his barn to get in to feed his cows. Snowdrifts are 8 m high.
Maritime provinces	1–2 Dec. 1964	Gales reach gust speeds of 160 km/hr. Three fishing boats, including two large draggers, are lost in the storm, accounting for 23 deaths and 8 reported injuries.
Southern Prairies	15 Dec. 1964	−34°C; 90 km/hr winds. Three people freeze to death and thousands of animals perish. Dubbed 'The Great Blizzard'
Montreal	7 Nov. 1969	60 hours; 70 cm of snow; 15 fatalities
Edmonton	1 Jan. 1973	A 707 aircraft crashes on approach to Edmonton International Airport and loses its crew. Highways are blocked, and the loss of power and communications in some rural areas makes survival difficult.
Southern Ontario	26–27 Jan. 1978	Record-breaking wind gusts and heavy snow cause 9 deaths and over $41 million in damages.
Iqaluit, NT	8 Feb. 1979	10 days; 100 km/hr winds; −40°C temperatures
Prince Edward Island	2 Feb. 1982	Islanders are marooned for 5 days in a crippling blizzard; winds of 80 km/hr whip a 60 cm snowfall into 7 m drifts. Succession of severe storms isolates many communities for days.
Southern Alberta	14–15 May 1986	A 2-day storm, described as the worst spring storm in living memory in Alberta, brings knee-deep snow and 80 km/hr winds.
Maritimes	1–4 Feb. 1992	A 110-hour-long blizzard with gusts up to 140 km/hr brings record snowfall of 163 cm to Moncton, NB; several other cities record more than 100 cm of snow; traffic is brought to a standstill; many stranded motorists have to be rescued; and many schools and businesses throughout the Maritimes are closed in the aftermath.
Trois-Rivières, Quebec	27 Feb.1992	Blowing snow produces near-zero visibility during an intense winter storm, causing a massive highway pileup involving 27 cars, 2 ten-wheelers, and 4 tractor-trailers. Two die and 15 are injured.

Table 3.2	Continued	
Location	**Date**	**Conditions**
Eastern Canada	15 March 1993	March blizzards caused by a mid-latitude cyclone that moved up the North American eastern seaboard eventually claim more than 240 lives. At one point over 3 million people are left without electricity. The storm causes the 177 m Liberian-registered freighter *Gold Bond Conveyor* to sink. Loaded with gypsum, it sinks about 200 km off Cape Sable Island in waves up to 20 m and winds gusting up to 120 km/hr, after setting sail from Halifax despite warnings of hurricane-force winds. Most crew members are from Hong Kong; none of the 33 aboard survive. Estimated cost: $19,866,000
Calgary	18 Dec. 1999	A howling blizzard strands more than 50 skiers at the top of Canada Olympic Park. Winds gusting to 70 km/hr force officials to close the hill. Some skiers are blown backwards by the strong winds. Others lie on the frozen ground, huddling together, until park staff send vehicles to bring them down.
St John's, NL	26 Feb. 2006	A massive blizzard buries eastern Newfoundland under up to 60 cm of snow. Winds gust to 130 km/hr and create snowdrifts up to 1.5 m deep, leaving many roads in the capital littered with stranded cars.

evidence that since the 1960s blizzards have become a less common occurrence in the western Prairies. The results of Lawson's study are consistent with an anticipated reduction in the number of Northern Hemisphere mid-latitude cyclones with global warming, and they may be an early indicator of such change.

Blizzards are one of the main meteorological events in the Canadian Arctic winter. Mean winds across a significant portion of the Canadian Arctic are generally stronger than those experienced over southern Canada. Their strength is a result of the strong pressure gradient established over the region. These persistently strong winds lead to a high frequency of blizzard conditions. Blizzards in the Arctic can be broken down into two types: the clear-sky, pure, blowing-snow type and the snow and blowing-snow type. The former are mainly restricted to the area north of the treeline, while the latter are the type commonly seen on the Prairies (Stewart et al. 1995). The significance of the difference between the two types is the fact that new snow doesn't have to fall for strong winds to create blizzard conditions; they can do so by entraining even a minimal amount of snow that is already on the ground. There is considerable variability in such conditions across the Arctic, with locations in the southwestern Arctic lacking the wind strengths experienced in the north and central Arctic (see Figure 3.4). This is a result of a dominant high-pressure system that extends from the Beaufort Sea along the Mackenzie Valley and into the Prairies. At the same time, there is routinely an area of low pressure over central or eastern Nunavut. The result is a northwesterly wind regime to the east of the ridge that can be strong enough to generate blowing snow.

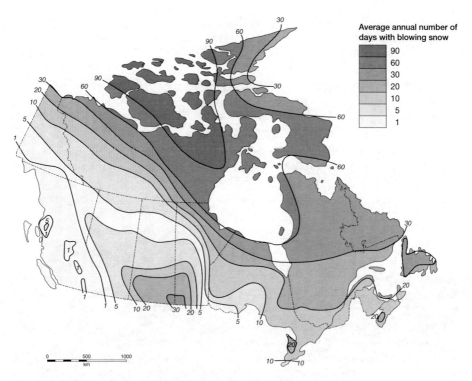

Figure 3.10 Average annual number of days with blowing snow in Canada.
Source: Modified from Environment Canada 1987.

Notable Canadian Blizzards

The blizzard that raged for 36 hours through 10–12 March 1888 would come to be known as the 'Blizzard of '88', or the 'Great White Hurricane'. The storm began as an unremarkable low-pressure system that originated in the Gulf of Mexico and travelled up the eastern seaboard. Although the storm occurred over one hundred years ago, it is still used as the benchmark against which current storms are often compared. Estimates indicate that property damage was in the tens of millions of dollars, and more than 400 deaths were reported in areas across northeastern North America (Kocin 1983). In comparing the Blizzard of '88 with other intense east coast storms, Kocin (1983) found that this system differed in several respects. In most cases, winter cyclones are preceded by an outbreak of very cold air across the eastern region of North America. There was no such air mass established over the mid-Atlantic states or New England before this blizzard. As a result, precipitation initially fell as rain and later changed to snow over the northeast. The stationary front and a large temperature contrast to the north of the system were additional unique features of this particular blizzard.

Between 31 January and 8 February 1947 in Saskatchewan, all highways into Regina were blocked as a result of a massive blizzard. Railway officials declared the conditions brought by this blizzard the worst in all of Canadian rail history. One train was buried in a snowdrift 1 km long and 8 m deep. Children stepped over power lines on their way to school and built tunnels to get to the outhouse. A Moose Jaw farmer had to cut a hole in the roof of his barn to get in to feed his cows (Environment Canada 2004i).

In 1982, a huge snowstorm, with up to 60 cm of snow, 100 km/hr winds, zero visibility, and wind chills of −35°C, paralyzed Prince Edward Island for a week. The storm buried vehicles, snowplows, and trains in 5–7 m drifts and cut off all ties with the mainland (Environment Canada 2004i). In Toronto in 1944, amid the news of the war, the headline 'Canada suffers worst blizzard in 72 years' made the news even in the United States. This severe winter storm dumped 48 cm of snow on the downtown area, while gale-force winds piled the snow into huge drifts. A total of 57.2 cm fell over two days. In all, 21 people died, 13 from overexertion. Funerals were postponed, expectant mothers walked to hospitals, and there were no home deliveries of milk, ice, or fuel. Of major concern, factories producing war material and equipment had to close temporarily (Environment Canada 2004i).

There have been a number of so-called Great Blizzards, at least three so-named in Alberta (1964, 1973, and 1989). The blizzards that occurred in southern Alberta between 17–20 and 27–29 April 1967 were certainly among the worst. A series of intense winter storms dropped a record of 175 cm of snow. Thousands of cattle, unable to forage for food in the deep snow, perished on the open range. Army units were dispatched to assist in snow clearing, while shipments of food, fuel, and feed were airlifted into the province. The good news? The revenue minister announced that the income tax deadline for residents of southern Alberta would be extended by two weeks, to May 15 (Environment Canada 2004i).

Another so-called Great Blizzard occurred between 12 and 15 March 1993. This one has also been referred to as 'The Storm of the Century' (Bowyer 2007) and was unique because of its shear size. It extended from Canada to Central America, but its main impact was on the eastern United States and Cuba. Forecasters cited formation factors that are likely to occur only once every 500 years. A disorganized area of low pressure formed in the Gulf of Mexico, joined an arctic high-pressure system in the American Midwest, and was brought into the mid-latitudes by a steep southward jet stream. For the first time, every major airport on the east coast, from Halifax to Atlanta, Georgia, was closed down for a period of time. Although the storm had a greater impact in the United States than in Canada (over 200 people died there), three storm-related deaths were reported in Quebec and one in Ontario. About 180 km south of Cape Sable Island, Nova Scotia, a 177 m ship sank in heavy seas, with all 33 of its crew lost at sea. Maximum waves were over 30 m high. Wind gusts of 233 km/hr were reported at Grand Etang, Nova Scotia, and were responsible for blowing the roof off the Cheticamp hospital (Bowyer 1995, 2007).

Ice Storms

Freezing rain is common in many parts of Canada. Victoria may have an average of only two hours of freezing rain per year, but Yellowknife, Regina, and Toronto endure

Figure 3.11 Blizzard of 1947.
Source: Journal Sentinel. Reproduced with permission of the publisher.

35 hours in the course of an average year and St John's, Newfoundland, a formidable 148 hours. Figure 3.12 illustrates is a wide spatial variation in trends across North America; central and eastern regions, for example, receive far more freezing rain and freezing drizzle than areas to the west. Episodes of freezing rain are generally short-lived, but from time to time they can develop into major ice storms that are notable both for their sparkling beauty and for the crushing weight of ice they leave on power lines and trees.

Damaging ice storms have occurred as recently as 1986 in Ottawa and 1984 in St John's, but the storm that struck much of eastern Canada in January 1998 was especially remarkable for its persistence, its extent, and, ultimately, the magnitude of its destruction. Over a period of six days, up to 100 mm of freezing rain fell intermittently over an area that at one point extended from central Ontario to Prince Edward Island. Downed power lines left nearly 3 million people without electricity. The Montreal area was the worst affected. The storm was blamed for 25 deaths, and with estimates of damages in the range of $1–$2 billion, it was by far the costliest weather disaster to afflict Canada up to the end of the twentieth century (Francis and Hengeveld 1998).

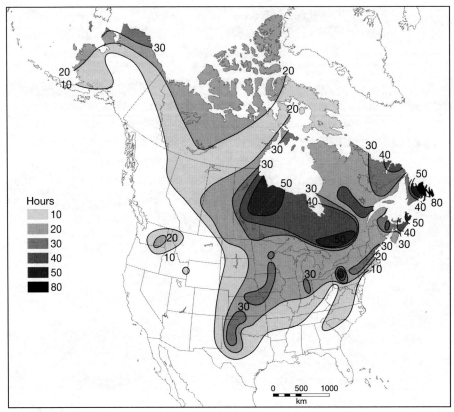

Figure 3.12 Median annual hours of freezing rain and freezing drizzle (combined), 1976–1990.
Source: Cathy Conrad.

Annual freezing rain averages range from 20–35 hours in the Prairies to 50–70 hours in the Ottawa Valley and southern Quebec. A drop in temperature to below 0°C can lead to a particularly dangerous situation. Surface temperatures hover above 0°C ahead of the cold front, and consequently, some of the initial snow melts on the surface or some of the initial precipitation falls as rain. The liquid on the ground later freezes in the cold air behind the front. This results in treacherous 'skating rink' conditions on surfaces, including sidewalks and highways (Stewart et al. 1995). On 19 January 2006, freezing rain left the city of Montreal in a treacherous mess. Heavy rain falling on a layer of ice made many streets and most sidewalks impassable. The ambulance service recorded 625 calls in less than 24 hours, with many pedestrians suffering from broken legs, wrists, arms, and head traumas. Rather than risk a long fall, some pedestrians opted to get down and crawl on their hands and knees (Sutherland 2006).

Freezing rain is defined as consisting of supercooled liquid water drops larger than 0.5 mm in diameter that freeze on impact with the ground or other objects. *Freezing drizzle* also freezes on impact with the ground, but it is composed exclusively of droplets less than 0.5 mm in diameter (Stuart and Isaac 1999). Freezing rain occurs when warm, moist air that is above the freezing point overlies cold air at ground level that is below the freezing point. As rain falling from the warmer air passes through the colder layer, it freezes instantly on contact with the ground. This process commonly occurs along fronts where warm and cold air masses on opposite sides of the freezing point meet, but the fronts generally move along quickly and thus the freezing rain conditions are short-lived. There are always unique exceptions to the norm, as was the case with the 1998 ice storm, which is discussed in more detail in the following section. That storm was the product of moist air out the of the American south, a steady flow out of the northeast that maintained a shallow layer of cold air in the lowlands of the Ottawa and St Lawrence River Valleys, and a stagnant ridge of high pressure over the Atlantic that kept the whole system in place.

Freezing rain and drizzle have been recorded at Meteorological Service of Canada (MSC) stations for many years. According to Stuart and Isaac (1999), freezing drizzle is more often observed than freezing rain virtually everywhere in Canada. Only in the Maritime provinces and St Lawrence River Valley are the occurrence frequencies of freezing rain higher or comparable to those of freezing drizzle. Atlantic Canada has some of the highest frequencies of freezing precipitation in the world—Newfoundland experiences more than 50 hours of freezing precipitation per year—while British Columbia and Yukon experience some of the lowest annual hours of freezing precipitation. Canada is never completely free of freezing rain, with totals in the summer months being confined to the far north. The severity of ice storms depends largely on the accumulation of ice, the duration of the event, and the location and extent of the area affected. Table 3.3 provides details of some of the more notable ice storms in recent Canadian history.

The 1998 Ice Storm

The ice storm that struck much of eastern Canada between 5 and 10 January 1998 had the largest ice accumulation, longest duration, and largest area affected of any ice storm in recent memory. The extent of the area affected by the ice was enormous (Figure 3.13). The area of freezing precipitation extended from Muskoka and Kitchener in Ontario through eastern Ontario, western Quebec, and the Eastern Townships to the Fundy coasts of New Brunswick and Nova Scotia. In the United States, ice coated northern New York and parts of New England (Environment Canada 2002b). The duration of this ice storm was particularly unique. On average, Ottawa and Montreal might receive between 45 and 65 hours of freezing precipitation per year. Over the course of five days, the ice storm of 1998 nearly doubled the annual total, with an excess of 80 hours (Environment Canada 2002b). Striking one of the most heavily populated parts of Canada, the storm directly affected more people than any previous weather event in Canadian history had.

Freezing rain started on Monday, 5 January 1998, as Canadians were returning to work after the Christmas holidays. For five days the freezing rain fell, coating Ontario, Quebec, and New Brunswick with 7–11 cm of ice. Trees and hydro wires fell, and utility poles and transmission towers came down, causing massive power outages, some lasting for as long as a month. Into the third week following the onset of the storm, more than 700,000 people were still without electricity.

Table 3.3 Significant freezing rain events in recent Canadian history.
Sources: Modified and compiled from Phillips 1990; Brun et al. 1997; Environment Canada 2004i, 2005f; PSC 2006a.

Location	Date	Conditions
Eastern Ontario	28–30 Dec. 1942	Ice as 'thick as a person's wrist' covers telephone wires, trees, and railway tracks. In Ottawa, 50,000 workers walk to work for 5 days. Because of the war, few men are available to clear streets and repair lines.
St Johns, NL	March 1958	43 hours of continuous freezing rain
Montreal	25 Feb. 1961	Ice storm heavily loads utility wires, causing them to snap. Some areas are without electricity for a week.
Southern Ontario	Jan. 1968	A 3-day storm with on-and-off rain and wet snow causes widespread power failure, closes schools, cancels food deliveries, disrupts mail and fire services, collapses buildings and antennae, isolates hospitals, and blocks highways.
Prairie provinces	6 March 1983	Most damaging in southern Manitoba. The storm forces Winnipeg International Airport to close for 2 days and topples several large TV towers. The freezing rain also causes other extensive damage.
Southeastern Newfoundland	11–13 April 1984	200,000 residents are left without power after an ice storm blankets transmission lines with 15 mm of ice, causing them to snap. The ice storm covers all of southeastern Newfoundland with 25 mm of glaze.
Ottawa	24 Dec. 1986	14 hours of freezing rain leaves one in 4 homes without power.
Eastern Canada	5–10 Jan. 1998	One of the most destructive and disruptive storms in Canadian history hits eastern Canada, creating hardship for 4 million people and costing $3 billion. Losses include millions of trees, 130 transmission towers, and 120,000 km of power and telephone lines. Power outages last several hours to 4 weeks. At least 25 deaths (mostly due to hypothermia) are attributed to the storm.
Winnipeg	1 Feb. 1999	A thick layer of ice forms on streets, causing dozens of car accidents; 850 claims are processed over 8 hours—the busiest day of the winter.
Calgary	30 Sept. 1999	Black ice conditions lead to 90-car pileup, closing Deerfoot Trail for 20 hours. It is so icy that paramedics have to abandon their vehicles and walk to the injured.
New Brunswick	2 Feb. 2003	Freezing rain causes thousands of power outages; it is estimated that the storm cost New Brunswick Power between $3 and $6 million in damage repair. A power company spokesperson calls the ice storm the worst in the utility's history, eclipsing even the 1998 ice storm in New Brunswick.

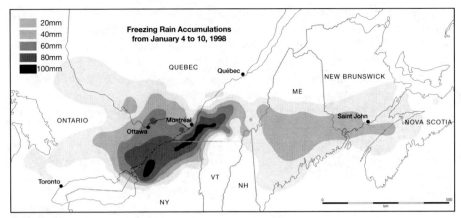

Figure 3.13 Freezing rain accumulations during the 1998 ice storm.
Source: Modified from Lecomte, Pang, and Russell 1998.

The damage in eastern Ontario and southern Quebec was so severe that major rebuilding, not repairing, of the electrical grid had to be undertaken. What had taken human beings a half-century to construct, it took nature a matter of hours to knock down.

> Farmers were especially hard hit. Dairy and hog farmers were left without power, frantically sharing generators to run milking machines and to care for new-born piglets. Many Quebec maple syrup producers, who account for 70% of the world supply, were ruined with much of their sugar bush permanently destroyed. (Environment Canada 2002b)

Cold and Wind Chill

Most Canadians have experienced the extreme cold of winter, and we have learned to adapt and modify our lives according to such extremes. In Fort Saskatchewan, Alberta, a bylaw was enacted to prevent people from leaving keys in the ignition of a running car. The cold weather had tempted people to leave their cars running during short stops, but that practice had also tempted thieves, who helped themselves to 35 such vehicles in a single week in January 1998 (Environment Canada 2004i).

Cold weather superlatives abound in Canada. Of the cities in the world with populations of 500,000 or more, Winnipeg has the coldest midwinter temperature. Ottawa has the distinction of being the second-coldest national capital, next to Ulaanbaatar, Mongolia. And Canada is the coldest country in the world. If you average all the daily temperature observations year-round, you get a chilling −5.6°C. Each year in Canada, more than 80 people die from overexposure to the cold and many more suffer injuries from hypothermia and frostbite (Environment Canada 2003a).

Every region in Canada except Prince Edward Island has experienced temperatures below −40°C (Phillips 1990). The lowest recorded Canadian temperature is −65°C, recorded in Snag, Yukon, on 3 February 1947 (the world record is −89°C, recorded in Vostok, Antarctica, in 1983). It was so cold in Snag that day that an exhaled breath made a hissing sound as it froze (*The Canadian Encyclopedia* 2007). A severe cold wave that extended from Alberta through to the Maritimes between January and February of 1994 provided some dramatic examples of cold extremes. At −40°C, tires were falling off their rims in Kapuskasing, Ontario. The case can certainly be made, however, that with global warming and climate change, we might not experience such dramatic lows in the relatively near future (Bonsal et al. 2001).

One of the most significant aspects of Canadian winters is the variability in daily conditions. Within Ontario alone, average daily temperatures can differ by as much as 20°C, with January temperatures in northern portions of the province often reaching −24°C, while southern regions might expect temperatures in the vicinity of −4°C (Environment Canada 2007k).

Areas that have extensive and frequent blizzard conditions are generally the areas most prone to very high wind chills. The combination of cold temperatures and strong northwesterly winds in such areas creates wind chills that are significantly higher than elsewhere in Canada. For example, in Baker Lake, Nunavut, on average, almost 80 per cent of the days in January and February have at least one report of a wind chill greater than 2200 W/m^{-2} (Watts per square metre) (Stewart et al. 1995).

Extreme cold temperatures are associated with continental arctic air masses (Brun et al. 1997), but the actual temperature reached depends specifically on the nature of the cold air mass and where it originated. Although cold temperatures are dangerous in their own right, they become more so in conjunction with strong winds. The combination produces a wind chill factor—heat loss measured in W/m^{-2}.

Anyone who has ever waited at a bus stop or taken a walk on a blustery winter day knows that we feel colder when the wind blows. On a calm day, our bodies insulate us somewhat from the outside temperature by warming up a thin layer of air close to our skin, known as the boundary layer. When the wind blows, it takes this protective layer away, exposing our skin to the outside air. It takes energy for our bodies to warm up a new layer, and if each one keeps getting blown away, our skin temperature will drop and we will feel colder (Environment Canada 2003a).

A 1999 public opinion research study conducted by the Meteorological Service of Canada indicated that 82 per cent of Canadians use wind chill information to decide how to dress before going outside in the winter. Many groups and organizations also use the system to regulate their outdoor activities. Schools use wind chill information to decide whether it is safe for children to go outdoors at recess. Hockey clubs cancel outdoor practices when the wind chill is too cold. People who work outside for a living, such as construction workers and ski-lift operators, are required to take indoor breaks to warm up when wind chill conditions are severe.

In April 2000, Environment Canada held the first global Internet workshop on wind chill. There were more than 400 participants from 35 countries, and almost all those attending agreed on the need for an internationally recognized standard method for measuring and reporting wind chill, one that would be more accurate, would be easy to understand, and would incorporate recent advances in scientific knowledge. In 2001, a team of scientists and medical experts from Canada and the United States worked together to develop a new wind chill index. The research branch of the Canadian Department of National Defence, already familiar with the issue because of its knowledge about how troops are affected by cold weather, contributed to the effort by conducting experiments using human volunteers.

The new index is based on the heat loss from the face (the part of the body that is most exposed to severe winter weather). Volunteers were exposed to a variety of temperatures and wind speeds inside a refrigerated wind tunnel. They were dressed in winter clothing, with only their faces exposed directly to the cold. To simulate other factors affecting heat loss, they also walked on treadmills and were tested with both dry and wet faces. Since the wind chill index represents the *feeling* of cold on the skin, it is not actually a real temperature, and thus it is given without the degree sign. For example, a meteorologist might say, 'Today the temperature is –10°C, and the wind chill is –20.' The new index is being used in both Canada and the United States, so travellers hear consistent information in the two countries (although in the United States the index is provided on a Fahrenheit scale) (Environment Canada 2003a).

'Extreme cold weather alerts' are issued by the city of Toronto's Hostel Services Unit when extremely cold weather will be or is currently forecast. Because severe cold can be a risk to homeless people, an alert is issued when (1) a low of –15°C is forecast, (2) the wind chill reaches the level at which Environment Canada issues a warning for outdoor activity, or (3) there are or will be extreme weather conditions, such as a blizzard or ice storm. In Canada, exposure to extreme cold claims more lives directly than any other atmospheric extreme. Table 3.4 provides details about some notable cold snaps. It should be noted that severe cold can occur well before the official winter season begins.

Many Canadians believe or have heard that the coldest corner in Canada is at Portage Avenue and Main Street in downtown Winnipeg. In fact, there are no official temperature measurements at any street corner in Canada to confirm the coldest intersection. Cold though Portage and Main may be, it is likely not the coldest spot in the Winnipeg area, as the nearby airport is usually 3°C–4°C colder than the downtown area. The lowest reading at the airport was –45.0°C on 18 February 1966, but airports in Edmonton, Regina, and Saskatoon have all recorded lower readings (*Canadian Encyclopedia* 2007).

Another cold Canadian weather myth is that White River, Ontario, is the coldest place in Canada. Its reputation for being the coldest area is probably based on the fact that for many years White River was identified as 'the coldest in the nation today' in the handful of newspapers and radio stations that reported daily temperature extremes.

Many Canadians are concerned with extreme cold because it affects their livelihood. Growers of perennial fruits in Ontario, for example, are particularly interested in extreme weather events and their effect on plant survival, growth, and development. There is relatively little that can be done, beyond choosing a good site, to avoid the negative impacts of cold temperatures once an orchard or vineyard is established. The difference between the temperature where no winter injury to orchards is observed and the temperature where severe injury is observed can be as little as 4°C (e.g., –26°C versus –30°C) (Ontario Ministry of Agriculture 2007).

Frost is another feature of severe weather that can have a devastating impact on crops. The extent of damage caused by frost depends on the temperature, length of

Table 3.4 Some notable Canadian cold snaps in recent history.
Sources: Modified and compiled from Etkin and Maarouf 1995; Brun et al. 1997; Environment Canada 2004i, 2005f; PSC 2006a.

Location	Date	Conditions
Ontario	29 Dec. 1933	Coldest date on record for many locations, including Ottawa at −38.9°C and Algonquin Park at −44°C. Records are also set in Manitoba, Quebec, and Nova Scotia.
Eastern North America	Feb. 1934	A cold wave engulfs the continent from Manitoba to the Atlantic seaboard. Ice traps fishing vessels off Nova Scotia, hospitals are jammed with frostbite victims, and for only the second time in recorded history, Lake Ontario freezes over completely.
Snag, Yukon	3 Feb. 1947	Lowest recorded Canadian temperature (−63°C)
Sisson Dam, NB	1 Feb. 1955	Coldest recorded temperature in New Brunswick (−47.2°C)
Red Deer	15 Dec. 1964	Wind chill temperature drops to −73°C. Over 1000 livestock perish, and 3 people freeze to death when the stoves in their homes go out at night.
Winnipeg	24 Jan. 1966	Longest skin-freezing wind chill (170 hours)
Edmonton	7 Jan.–2 Feb. 1969	Temperatures below −18°C for 26 days
Lethbridge	15 Jan. 1971	Most rapid winter temperature increase (from −20°C to 1°C) in the span of an hour
All of Canada	1972	One cold year! The only year on record when all weather-reporting stations in Canada reported temperatures below normal on an annual basis.
Pelly Bay, NT	13 Jan. 1975	Coldest recorded wind chill (equivalent to −97°C)
Saskatoon	28 Dec. 1978	Longest wind chill event (215 hours)
Yukon to Ontario	8–20 Feb. 1979	Weather slows TransCanada and Alaska Pipelines oil flow to one-fourth its normal speed; furnaces break down from lack of oil. Pipes burst across Metro Toronto. On 20 February 1979, for the first time in recorded history, all 5 Great Lakes freeze over, stopping all water traffic. In a snowstorm in Iqaluit, NT, temperatures drop to −40°C and winds reach 100 km/hr; snow keeps residents indoors for 10 days.
Most of Canada	Jan. 1982	Coldest winter month on record; most provinces below −40°C. Three-week cold spell; trucks and trains fail and are abandoned until milder weather comes; a damaged steel bridge forces a 200-km detour off the Alaska Highway; more than 25 Ontario highways are closed due to blowing snow and poor visibility. In the northern Prairies, temperatures as low as −47°C are recorded on 17 Jan.
British Columbia	30 Jan. 1989	Freezing causes $2.5 million in damage.

Table 3.4 Continued

Location	Date	Conditions
Edmonton	29 Jan. 1989	Most rapid temperature drop (from 2°C to −12°C) in the span of an hour
Yellowknife	31 Dec.–19 Jan. 1993	A record 20 consecutive days with temperatures below or equal to −37°C

exposure, and humidity levels, although the potential damage is hard to predict given the variety of factors that affect the tolerance of plants. In general, a −2°C to −3°C frost over a period of at least an hour can be expected to cause damage to crops, and a −1°C frost for an extended period (three to four hours) can cause similar damage (Saskatchewan Agriculture 2007).

So Cold That Niagara Falls Freezes?

This is one myth that is actually true. The flow of water over both the American and Horseshoe falls stopped completely on 19 March 1848—the only time such a phenomenon is said to have happened (Figure 3.14). The falls did not actually freeze over, but the flow stopped to the point where the pre-existing water was able to freeze. Despite the limited and slow communication networks of the day, a reported 5000 people from as far away as Hamilton, Ontario, and Buffalo, New York, converged on the scene, jamming local roads. Thousands attended special church services, convinced that the incident was a sign of greater disasters to come (Heidorn 2000).

Figure 3.14 Niagara Falls freezes.

There was nothing particularly unusual about the winter of 1847–8, nor of the onset of spring in March. The cause of the water stoppage appears to have been an ice jam that formed at the source of the Niagara River. Normally, in late March, the ice that has covered Lake Erie during the winter begins to break up, and the larger chunks are pushed by the wind and currents towards the eastern end of the lake. There the chunks can pile up in ridged and rafted ice, but usually these walls of ice do not touch the riverbed. The freezing of Niagara in 1848 was so remarkable that decades later eyewitnesses to the event were asked to sign declarations swearing to their presence there (Heidorn 2000).

Visualizing the Cold

Most people don't think about the cold being visible, at least certainly not from space. But this is precisely what the Terra satellite recorded on 25 January 2004. Terra is a multinational, multidisciplinary mission involving partnerships with the aerospace agencies of the United States, Canada, and Japan. On 24 February 2000, Terra began collecting a 15-year global data set on which to base scientific investigations about the Earth. Together with the entire fleet of EOS (Earth Observing System) spacecraft, Terra is helping scientists better understand climate and environmental change.

Black Ice

Black ice is the term given to the thin layer of ice that can form on the surface of a road. Because of its transparent appearance, the ice takes on the colour of whatever surface it is overlaying. In most cases, the surface is asphalt pavement, which makes the ice appear black—an obvious hazard to unsuspecting drivers. Black ice has been known to form during periods of light rain or drizzle and where road surface temperatures have fallen below 0°C (AMS 2008). Since black ice forms mostly on surfaces where the temperature has fallen below 0°C, the surfaces of roadways, bridges, buildings, and even maritime vessels are especially susceptible, since their temperatures can drop faster than the surrounding surfaces. Because of the hazard potential, many provinces and municipalities take preventive measures when the weather forecast calls for potential freezing/freezing rain conditions. The expenditures in terms of time and resources for road salt and sand can be considerable. In Ontario alone, between 500,000 and 600,000 tons of road salt are spread each year to prevent slippery road conditions (Ministry of Ontario).

On 7 February 2008, in a small community on Vancouver Island, a bizarre and rare accident was triggered by black ice conditions. Emergency crews were responding to a car crash, but when two fire trucks ploughed into an ambulance shortly after it had arrived, they quickly became part of the crash scene itself. A section of the Trans-Canada Highway was closed for several hours while tow trucks hauled away emergency vehicles as well as a tractor-trailer that had also become entangled in the pileup (*Province* 2008).

Winter Weather and a Changing Climate

The winter of 2006–7 is tied with the winter of 1986–7 as the second-warmest winter since records began to be kept in 1948 (2005–6 was the warmest). Temperatures were 3°C above normal. Almost all of Canada experienced above normal temperatures in the winter of 2006–7, with northern British Columbia, eastern Yukon, and the eastern Northwest Territories all experiencing temperatures greater than 5°C above normal (Environment Canada 2007k). Winter temperatures in Canada have generally been increasing nationally, with a warming trend of 2.3°C from 1948 to 2007 (Figure 3.15).

Throughout the twentieth century, most of southern Canada experienced fewer days with extreme low temperatures during the winter, spring, and summer months (Bonsal et al. 2001). The Intergovernmental Panel on Climate Change (IPCC 2007b) has reported that, globally, cold days, cold nights, and frost have become less frequent. Hanesiak and Wang (2005) conclude that freezing rain is generally increasing in frequency in northern Canada, while blowing-snow events are generally decreasing in frequency and fog occurrences are showing regionally varying trends. Zhang et al. (2001) concludes that although heavy snowfall events in southern regions of Canada showed an upward trend from the beginning of the twentieth century through the late 1950s to 1970s, there has since been a consistent downward trend to the present. Conversely, heavy snowfall events in northern regions of Canada have become more frequent in the last 50 years.

Since the mid-1980s, many parts of Canada have been receiving smaller amounts of snowfall. One might conclude from this that severe storms have been on the decline in these regions, but the evidence to support this is mixed. Lambert (1995,

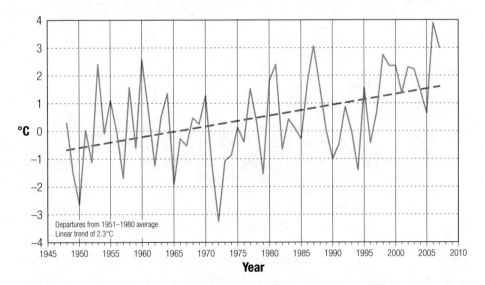

Figure 3.15 Winter national temperature departures and long-term trend, 1948–2007.

Source: Modified from Environment Canada 2007k. © Her Majesty the Queen in Right of Canada, 2007. Reproduced with the permission of the Minister of Public Works and Government Services Canada.

2004) has examined winter storm activity in the extratropical Atlantic and Pacific that has occurred since the beginning of the twentieth century, and he concludes that there was little change in the number of low-pressure systems (the marker for unusually severe storms) in the years before 1970. After 1970, however, severe winter storms have shown an increased frequency, particularly in the Pacific. In a study of the Canadian Prairies, however, Lawson (2003b) noted that although blizzard events on the Prairies tend to be diminishing, those that do occur show increasing—not decreasing—trends in their severity. The author notes that there is a need to devote further study to the impacts of climate change at regional spatial scales. Not only will snowfall amounts and storm frequencies be modified by a changing climate, but extreme temperatures will also be altered. Lawson (2003a) shows a trend towards fewer winter extreme cold temperatures on the Canadian Prairies and concludes that the results are 'consistent with the expectations of enhanced greenhouse warming' (238).

Between 1999 and 2004, the Canadian Climate Impacts Scenarios (CCIS) Project provided bioclimate profiles for 112 stations distributed across Canada. Bioclimate profiles are based on daily observations—taken over as many or as few years as desired—for maximum and minimum temperature, total precipitation, total rain, total snow (expressed as water equivalent), and total snow depth. The daily station data used in the study were from the National Data Archive and were provided by the Meteorological Service of Canada. The graphs in Figure 3.16 show the potential changes in total amounts of precipitation between 1951–90 and 2070–99 in Prince George, BC; Alert, Nunavut; and St John's, Newfoundland. The amount of precipitation that falls in the form of snow is reduced in all three locations, with an annual decline of 47, 39.4, and 54.6 cm respectively per year. The decline in total precipitation falling in the form of snow is compensated for by the increased amount of annual rainfall, with an average increase of 32 mm of annual precipitation for the three locations.

Looking Ahead

Canadians are used to severe winter weather. We adapt our lives accordingly while we await the warmer seasons. From a scientific perspective, there is much about the weather that remains poorly understood. For example, downslope winds, orographically affected storms, snow bands, blowing snow, and blizzards need to be studied in much more detail. The challenge of the future is to gain an understanding of all aspects of winter storms in Canada. This can only be achieved through a collaborative effort involving the operational forecasting community as well as the research communities of atmospheric and social scientists. Because winter weather, in all its forms, has such a tremendous impact on the lives of Canadians, we must jointly tackle the scientific and socio-economic challenges posed by the ubiquitous features of our climate (Stewart et al. 1995).

Severe 'Winter' Weather   75

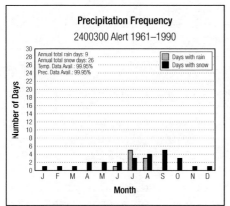
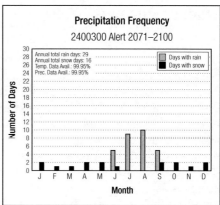
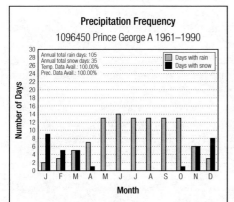
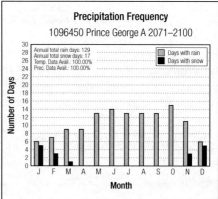
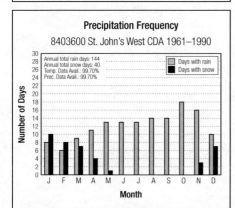
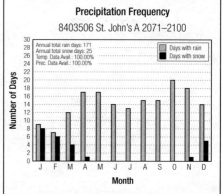

Figure 3.16 Canadian Climate Change Scenarios Network Bioclimate Profiles (predicted change in precipitation form and frequency from 1961–1990 to 2071–2100) for Alert, Prince George, and St. John's (using an ensemble mean scenario).

Source: Modified from http://www.cccsn.ca/Scenarios/BioclimateTool/Bioclimate-e.phtml.

Chapter Summary

Canada is associated with heavy winter weather in the minds of Canadians and non-Canadians alike. The most common cause of severe winter weather in Canada is the extratropical cyclone, a weather system that can produce heavy snowfall, blizzards, freezing rain, and severe cold weather. Winter low-pressure storm systems favour certain tracks, along which broad areas of cloud cover, snowfall, rain, high winds, and sometimes freezing precipitation will be found. A variety of strong wind systems affect regions of Canada at varying spatial and temporal scales through the year, including so-called Wreckhouse winds, chinooks, Alberta clippers, and nor'easters. There is a substantial variation in the patterns of annual snowfall across Canada, although generally the largest amounts are in the northwestern Pacific coast, high altitudes in British Columbia, Ontario northeast of Lake Superior, and the higher elevations of Quebec and Labrador and Newfoundland. There does not need to be a lot of snow on the ground for there to be drifting and blowing snow, which explains why regions that don't have large snow accumulations, such as northern Yukon and Nunavut, can have high frequencies of blizzard conditions. Areas that have extensive and frequent blizzard conditions also tend to be the most prone to very high wind chills. There are wide spatial variations in freezing precipitation trends across North America, although central and eastern regions receive far more freezing rain and freezing drizzle than areas to the west. Winter temperatures in Canada have generally been increasing nationally, with a warming trend of 2.3°C over the last 60 years. Climate change will have an impact on snowfall amounts and storm frequencies in Canada.

Chapter Four

Severe 'Summer' Weather

Objectives

- To provide an overview of the variety of severe summer weather conditions that exist across Canada
- To introduce processes associated with severe summer weather in Canada and the storm tracks they tend to follow
- To explain and describe heat waves in Canada and to introduce the concept of the urban heat island effect
- To describe humidity and smog conditions across Canada
- To describe the spatial trends and characteristics of thunderstorms, tornadoes, hail, and lightning across Canada
- To discuss the implications of climate change for severe summer weather in Canada

Introduction

Canada doesn't immediately come to mind when one thinks of severe summer weather. Mention tornadoes and people usually think of 'Tornado Alley' in the United States. Mention a heat wave and we might think of lower latitudes than our own. Perceptions aside, Canada has its fair share of summer-related severe weather, including heat waves and high humidity, thunderstorms, hail, tornadoes, and lightning. In 2007, residents of Alberta, Saskatchewan, and Manitoba experienced a record number of tornadoes, hailstorms, and severe summer weather warnings, abnormal activity that Environment Canada named one of 'top ten weather stories' of the year (Environment Canada 2008a). Abundant spring rains followed by excessive heat and humidity and an active jet stream resulted in the perfect recipe for violent weather. The record number of weather events (410 in total) eclipsed the previous high of 297, set in 2006. Especially frequent were the hail events, setting record numbers for all three provinces. Although Canada might be best known for having severe winter storms, its summer weather conditions can also be severe.

Summer Storm Tracks

The individual processes that create the variety of severe summer weather conditions we experience at any one time in Canada are system-specific (e.g., extreme heat, thunderstorms, and tornadoes). Whole regions of the country, however, are affected by larger-scale weather patterns (Figure 4.1). The Prairie provinces, for example, are affected by the Mackenzie Low, the Alberta Low, and, to a lesser extent, the Colorado Low, all of which are loosely categorized on the basis of the region over which they develop and the circumstances of their formation. These low-pressure storm systems can occur at any time during the year but are more intense in the winter months, when the temperature differences between air masses are more pronounced. The storm systems associated with these lows in the summer months tend to produce showers and thundershowers. Clearing behind these lows tends to be more rapid in the summer than in the winter, when it is relatively gradual. In the summer, apart from the low-pressure systems tracking across the Prairies, the main concern in the region is convection. Under these circumstances, air rises along a frontal system as a result of the convergence of surface winds, leading to the formation of cumulus clouds and potentially the development of a thunderstorm. The most active region of convective weather

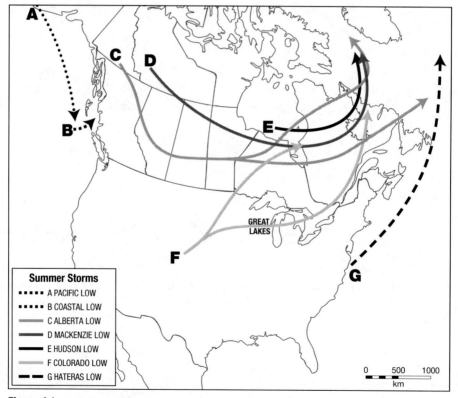

Figure 4.1 Summer storm tracks across Canada.

Source: Modified from Nav Canada.

conditions on the Prairies is located in southern Saskatchewan and central Alberta, with July being the most likely month for convection to occur.

On the west coast of Canada, storms are much fewer in number and far less severe. Low-pressure areas tend to remain offshore, and systems that do move through usually occur to the north. Minor frontal systems and thunderstorms generate most of the adverse weather conditions of regions to the south. In August and September, there can be weeks between weather systems (Nav Canada 2008). In Ontario, Quebec, and the Atlantic provinces, the frequency and severity of the storm systems tracking through the region are also generally reduced and low-pressure areas follow a more northerly track. This northward shift pushes the main storm track to the west and north of the Atlantic provinces, with minor frontal systems and thunderstorms producing most of the weather that otherwise passes across the region. Severe storms such as tropical cyclones are the exception; they are covered in detail in Chapter 6.

Heat

On 1 August 1 2006, temperatures in much of southern Ontario hit record highs. Peterborough broke its previous record with a high of 36.7°C, nearly three degrees higher than the record temperature for the same date in 1975. On the same day, Toronto broke its previous daytime high from 1955 when temperatures soared to 36.8°C; St Catharines hit a humidex of 49°C (CBC 2006h); parts of Ontario were under smog and humidex advisories; and a number of communities issued heat alerts. In Toronto, an extreme heat alert remained in effect for three days. Hamilton had the misfortune of having scrapped its heat alert program only a few days prior to the heat wave. City health officials had said that heat alerts don't work and that the city's temperatures weren't extreme enough to warrant them. As temperatures soared just days later, local politicians demanded that the system be brought back immediately, arguing that the alerts protect the most vulnerable members of the community (CBC 2006c). The system was reinstated but has since been criticized for lacking mitigation and preventive measures. Ontario was not the only province setting records. In Winnipeg, 16 days in July 2006 had temperatures above 30°C. Normally, there would be only four days with those temperatures (CBC 2006n). In British Columbia in July 2006, fifty-year temperature records were broken as a massive heat wave moved through greater Vancouver and parts of the BC interior (CBC 2006a).

Heat waves have not been viewed in the past as a serious threat in Canada (Smoyer-Tomic et al. 2003), although the impact of severe heat outbreaks can be substantial. In a climate-changing world, there is the valid concern that Canadians will increasingly face severe heat waves. Although the destruction from hurricanes, floods, and tornadoes may draw more attention, the death toll from severe heat waves can be substantially higher than that from the more infamous weather events. Increases in deaths associated with extreme heat waves have been well documented in southern Ontario and Quebec (Smoyer-Tomic et al. 2003). A heat wave can be thought of as simply a period of unusually hot weather, but according to Environment Canada's more scientific and absolute definition (1996), a *heat wave* is a period of more than three consecutive days of maximum temperatures at or above 32°C. The adverse effects of heat on humans, however, have been observed at less extreme temperatures and for shorter durations.

Heat waves have notable effects on human mortality, regional economies, and ecosystems. Two well-documented international examples are the Paris heat wave of August 2003 and the Chicago heat wave of 1995 (Meehl and Tebaldi 2004). Across central Europe, 2003 was the hottest summer on record since 1500 (Kovats and Ebi 2006). The Paris heat wave was considered responsible for the death of an estimated 14,802 individuals, the large majority of whom were elderly individuals over the age of 75 (Belmin et al. 2007). Heat waves in the United States in 1980 and 1988 resulted in an estimated 10,000 and 5,000–10,000 deaths respectively (Smoyer and Rainham 2000). In their study of five Canadian cities, Smoyer and Rainham (2000) observed that heat stress–related mortality varied among places with relatively similar weather conditions. Urban areas experience higher temperatures than less built-up areas, with evening temperatures remaining high, and this creates more dangerous conditions for human health because of the lack of respite from the high temperatures. Smoyer and Rainham concluded that as urban development continues along the Toronto–Windsor corridor, population vulnerability to heat stress is likely to increase. However, since heat stress conditions are predictable, and given a combination of hot weather warnings and watches, public education campaigns, and well-organized plans for dealing with heat emergencies, the harmful impacts of heat stress conditions on health can be mitigated. The public health measures that were implemented in many parts of Europe after the 2003 heat wave have centred almost exclusively on heat health-warning systems that use forecasts of high-risk weather conditions to trigger public warnings (Kovats and Ebi 2006). The mid-July 1995 Chicago heat wave led to approximately 600 heat-related deaths over a period of five days as temperatures soared to 41°C (Klinenberg 2002). The impact of this heat wave was aggravated by inadequate warnings, power failures, inadequate ambulance service and hospital facilities, and lack of preparedness (Duneier 2004).

Several characteristics of heat waves, such as their frequency, duration, and intensity, affect their impact. If cooling occurs in the evening to offer relief from daily highs, their effects are not as severe. In areas where hot and humid summer conditions are common, physiological, behavioural, and infrastructural adaptations are likely to have taken place, reducing the harmful effects of heat stress (Smoyer-Tomic et al. 2003). In areas where extremely hot weather is relatively uncommon, such adaptations are less likely to have been made, leaving populations more vulnerable to harmful effects. No population, however, is completely immune to very hot weather, and studies generally indicate that mortality increases at extreme temperatures (Kovats and Ebi 2006).

In their study of heat wave impacts in Canada, Smoyer-Tomic et al. (2003) note that heat stress events have affected most areas of Canada to varying degrees. More events occur in Ontario and Quebec than elsewhere in the country, yet heat waves have had harmful effects in the Atlantic provinces as well as in Yukon. The Prairies, southern Ontario, and the St Lawrence River Valley region of Quebec have experienced the highest incidence of heat waves, while the North, the Pacific coast, and the Maritime provinces have experienced the lowest frequency. Toronto has the highest count of extreme heat events in Canada, averaging four episodes per year for the 62-year period of record. It is therefore not surprising that most heat wave research has focused on Toronto and southern Ontario.

Summer weather conditions vary across Canada, as do perceptions of their severity, which depend on the climate a person is used to. While a temperature of 30°C in Vancouver or Charlottetown would be highly anomalous and could exact a great toll on human health, the same temperature in Winnipeg, where summer temperatures above 30°C are more common, would be expected to have lesser health impacts. Similarly, people in the Arctic might be vulnerable to modest temperature increases, while people in southern Ontario might be considered somewhat acclimatized to summer heat waves. Clearly, absolute increases in temperature alone can be misleading, and relative increases should be kept in mind. Table 4.1 provides information about the significant heat waves that occurred in Canada between 1912 and 2006. There remains a need to conduct further research on the occurrence and effects of heat waves in Canada.

Table 4.1 Significant heat waves in Canada, 1912–2006.
Source: Compiled and modified from Canadian Disaster Database 2006; Smoyer-Tomic et al. 2003; Khandekar 2002; Environment Canada 1999, 2005d, 2006e.

Location	Date	Conditions
Eastern Canada	8–10 July 1912	Temperatures of >32°C from Ontario to the Atlantic Ocean. At least 3 heat-related deaths in adults are reported, while many more infant deaths are believed to have occurred in poor areas. One Ontario farmer commits suicide, and a mother kills her children (9 July). Wheat, barley, and oats grow to be below average in height and grade.
Across Canada	5–17 July 1936	The hottest month in Canada in the 20th century: from southern Saskatchewan to the Ottawa Valley, temperatures exceed 32°C for one and a half weeks. St Albert, MB, reaches 44.4°C, and Atikokan, ON, hits 42.2°C. A total of 1180 people die across Canada. The heat is so intense that steel rail lines and bridge girders twist, sidewalks buckle, crops wilt, and fruit bakes on trees.
Midale and Yellowgrass, SK	5 July 1937	The temperature reaches at 45°C, the highest ever recorded in Canada.
Toronto, ON	24 Aug.–5 Sept. 1953	The longest heat wave on record. One person dies of heat-related fever, and 186 cases of heat prostration and injury are reported at the Canadian National Exhibition. Water consumption reaches record level of 182 million gal. for Toronto area on 2 September. Lake Erie is warmed to the point where millions of fish are killed.
Nova Scotia	26–28 July 1963	The longest sustained heat wave in Halifax since 1928, with temperatures of 33°C–34°C: greater traffic volume causes many minor accidents, and several people go to hospital; 4 individuals drown, and one dies in a burning car.
Grande Prairie, AB	28 May 1970	A heat wave with temperatures soaring to 40°C

Table 4.1 Continued		
Location	**Date**	**Conditions**
Prairie provinces to Ontario	5–11 July 1988	A heat wave in the Prairie provinces and central and southern Ontario results in 14 elderly people dying of heat-related factors. On 7 July, the high in Toronto is 37.2°C (hottest day since 1953); on 6 days, afternoon highs exceed 35°C; air pollution levels soar; power and water consumption reach record levels; on 28 July, rotating blackouts ease the power drain that has threatened to cause a complete power failure; ice companies are overwhelmed with orders and quickly sell out.
Edmonton	25 July 1994	Soaring temperatures lead Edmontonians to use 997 MW of power per household to light and cool their homes, setting a record for summertime energy consumption.
Quebec	Aug. 1995	Temperatures up to 6°C above normal are registered.
Atlantic Canada	Summer 1999	For third year in a row, extreme heat and drought wilt crops, endanger livestock, and prompt fire bans.
Southern Ontario and Quebec	Summer 1999	Twice as many hot days (>30°C) are experienced as in an average summer. Ottawa experiences 26 days >30°C compared to a norm of 12 days per year.
Northwest Territories	Summer 1999	Minimal precipitation and hot dry weather contribute to extreme fires in the Northwest Territories, eventually burning 3 million hectares of boreal forest.
Southern Ontario and Quebec	1–3 Aug. 2006	Just north of Toronto at Buttonville Airport, the temperature reaches 37.8°C on 1 August 2006. On the same day, the nighttime minimum temperature in Toronto is highest ever recorded, only dropping to 27.2°C. In Ottawa, the temperature reaches 36.3°C, but with the humidity factored in, it reaches an all-time humidex record of an oppressive 48°C. Record power consumption is recorded in Ontario when consumers use 27,000 MW.

The Urban Heat Island Effect

Heat islands—the phenomenon that describes urban and suburban temperatures that are 1°C –6°C hotter than those in nearby rural areas (EPA 2008)—can function to trap long-wave radiation (IR) emission from the relatively hot urban summer surfaces. When this radiation is reflected back towards the surface (as incoming long-wave radiation) after striking the urban or industrial atmosphere, the increase in surface energy (expressed as W/m–2) can lead to further surface warming. The first study to document this was conducted in Montreal (Oke and Fuggle 1972). Remarkably large heat island effects were subsequently documented in the heavily industrialized atmospheres of Hamilton and Windsor (Rouse et al. 1973; Brazel and Osborne 1976). The altered energy balance creates cities that have a tendency to be warmer than their surrounding environments.

When warm summer temperatures reach extreme levels, the situation can be exacerbated in the urban core by the heat contributed by buildings, cars, and the industrial production of heat. In winter, the values of heat produced by a central city and its suburban ring can equal or surpass the amount of heat available from the Sun. It is therefore not surprising that urban areas—in both summer and winter—are generally warmer than the adjacent countryside (de Blij et al. 2005). The 'urban heat island effect' has been documented in cities and towns from every region of Canada. This effect tends to be at its maximum about three to four hours after sunset, and at its minimum in the early afternoon. The effect is therefore a principally nocturnal phenomenon (Bailey et al. 1997). Factors determining the way in which an urban area is affected by oppressively hot weather include city location, heat island magnitude, and housing conditions (Sheridan and Kalkstein 1998). Spatial patterns of the heat island indicate a strong correlation between temperature increases and land use patterns and building density, with urban cores tending to have maximum temperature increases. The magnitude of the effect is influenced by weather conditions, as well as by the city size. For some of Canada's largest cities, the heat island effect on calm cloudless nights can cause temperatures to rise by as much as 12°C (Landsberg 1981).

The consequences of higher temperatures in urban areas can be significant. Cities located in the middle latitudes, where summer brings irregularly occurring but intense heat waves, demonstrate the strongest response to heat stress (Dolney and Sheridan 2006; Sheridan and Kalkstein 2004). In Canada, downtown Toronto experiences significantly more heat wave events than Pearson International Airport. Likewise, downtown Edmonton, Ottawa, and Montreal all experience more heat waves than their less-developed peripheries (Smoyer-Tomic et al. 2003). According to the records of Toronto Public Health, in the hot summer of 2005, over 12,000 people used the 'cooling centres' during the heat emergencies and 450 calls were made to the city's 'Heatline', which offered advice and recommended cooler locales (in extreme circumstances, ambulances were sent) (Sheridan 2006). As already noted, during the 2003 heat wave that extended across Europe, Paris was hit particularly hard. As temperatures rose to 40°C, an estimated 180 people in the city died in one day alone because of the abnormal heat.

Recent work on so-called green roofs (roofs with a vegetated surface and substrate) has proved that they are a promising mechanism to reduce the urban heat island effect. During warm weather, green roofs reduce the amount of heat transferred through the roof, lowering the energy demands of air-conditioning systems. In addition, the vegetated surfaces that replace dark and impervious surfaces act to increase the surface albedo (amount of solar radiation reflected back into the atmosphere). A study in Toronto indicated that if 50 per cent of the city's buildings had green roofs, the temperature reductions could be as great as 2°C in some areas (Bass et al. 2003).

Humidity

In the summer of 2007, high humidity on the Canadian Prairies resulted in a humidex reading of 53 in Carman, Manitoba. This broke the Canadian record, which had previously been 52.1, set in Windsor, Ontario, in June 1953. The controlling weather system—a huge upper ridge situated in the American Midwest—was a little farther east than normal, enough to tap moisture from the Gulf of Mexico and move it into the Prairies. Combined with local moisture sources from transpiring crops and evaporating surfaces soaked by significant spring rains, the system filled the air with insufferable humidity. From the 22nd to the 24th of July, Regina recorded 29 hours of humidex values above 40, including a reading of 48 for two consecutive hours. Without question, Regina was experiencing the most uncomfortable spell of weather in its history (Environment Canada 2008a).

There were several unexpected impacts from these extraordinarily high humidity levels. Hundreds of elective surgeries at several hospitals across the West had to be cancelled, because high humidity compromises sterile equipment and increases the risk of post-operative infection. The warm, moist air put a strain on utilities and helped establish a new summer record for power consumption, one that came close to the wintertime record peak load. In some areas, bloated fish floated onto the shores of several overheated lakes, streams, and reservoirs. 'Summer kill' occurs when conditions of high temperatures and little or no wind lead to oxygen depletion, suffocating fish. Hundreds of dead ducks also turned up in lakes east of Calgary, probable victims of a toxin that thrives in hot, dry weather (Environment Canada 2008a). On 20–21 July 2006, firefighters in Halifax responded to dozens of false alarms because of the hot, sticky weather. The high humidity had triggered many smoke alarms that could not distinguish between smoke and high water content in the air (CBC 2006e).

Humidity is a general term that refers to the concentration of water vapour in the atmosphere. *Relative humidity* is a measurement that indicates the amount of moisture that the air contains compared with the amount it is capable of holding at that temperature. A figure of 100 per cent would mean that the air is saturated. At this point, mist, fog, dew, and precipitation are likely (Environment Canada 2006c). Relative humidity is normally at its maximum when the temperature is at its lowest point of the day, usually at dawn. Even though the *absolute humidity* may remain the same throughout the day, the changing temperature causes the ratio to fluctuate. The *humidex*, a Canadian innovation first used in 1965, was devised to describe how hot, humid weather feels to the average person. It combines temperature and humidity into one number to reflect the perceived temperature. Because the humidex takes into account the two most important factors that affect summer comfort, it can more accurately convey how stifling the air actually feels than either factor can do alone.

The humidex is widely used in Canada, although extremely high readings are rare except in the southern regions of Ontario, Manitoba, and Quebec. Generally, the humidex decreases as latitude increases. The hot, humid air masses that cause this kind of uncomfortable weather usually originate in the Gulf of Mexico or the Caribbean. The humidex values and associated discomfort levels listed below are culturally and regionally dependent. People residing in areas with low average temperatures and humidity levels rarely above 30 would likely be very uncomfortable with a humidex of 39, for example.

- Less than 29: No discomfort
- 30 to 39: Some discomfort
- 40 to 45: Great discomfort; avoid exertion
- Above 45: Dangerous
- Above 54: Heatstroke imminent

Smog

Smog is a severe weather phenomenon that occurs in many parts of the world. In 2005, Ontario issued 53 smog alerts (Figure 4.2). In that province alone, smog accounts for over 1800 deaths annually (Green Ontario 2007), and there are estimates that this figure could increase to 10,000 (CTV 2006). The Ontario Medical Association has made 20-year projections of the consequences of smog for the health-care system, and according to these, deaths among seniors could increase by about 80 per cent.

The word *smog* is a combination of the words 'smoke' and 'fog' and refers to the brownish haze that can hang over cities. Ground-level ozone (as opposed to the ozone layer found in the high atmosphere) is the main constituent of urban smog. Smog formation and ground-level ozone buildup occur most often during hot and sunny afternoons in the summer months. A combination of solar radiation and volatile organic compounds (VOCs)—mainly produced by the evaporation of liquid fuels, solvents, and organic chemicals—react with nitrogen oxides (NO_x) produced in the

Figure 4.2 Smog in Toronto.
Source: Dick Hemingway Editorial Photographs.

combustion of gas, diesel fuel, and coal. The product of this reaction is ground-level ozone. Ozone and its precursors can be transported over large distances in the atmosphere by the winds (Environment Canada 2007d).

Other factors in the formation of smog include such local conditions as the size of cities, the density of population and traffic, and industrial activities. Smog exists not only in and near urban centres but also in rural regions, to which it is carried by winds. Typically, it has formed by mid-afternoon over urban centres and then spreads across rural regions late in the afternoon or in the evening. An ozone episode has a lifetime from a few hours to several days (Environment Canada 2007d). Canada's worst smog corridor extends from Windsor through to Montreal, although smog days do occur in many other regions of the country (Figure 4.3). Around the Bay of Fundy, in southern New Brunswick, and in parts of Nova Scotia, 50–80 per cent of the smog is the result of cross-border pollution from the northeastern United States and emissions from central Canada. As much as 80 per cent of the ground-level ozone and smog conditions in the Lower Fraser Valley in southern British Columbia (including Vancouver) comes from local sources, particularly tailpipe emissions (Government of Canada 2002). Smog season in Canada tends to last from May through September.

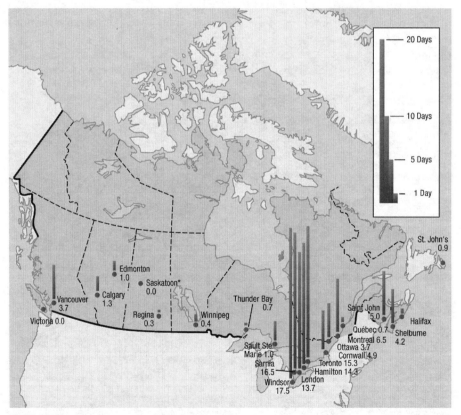

Figure 4.3 Average number of days (1987–1992) when one-hour average concentrations of ozone surpassed 82 parts per billion (ppb). An excess of 65 ppb is indicative of relatively poor air-quality conditions.
Source: Cathy Conrad.

The province of Quebec has one of the most comprehensive smog-monitoring and -warning systems in the country, offering smog forecasts to 95 per cent of the Quebec population. 'Info-Smog' runs year-round, with a program for winter (1 November until 15 April) since 2001 and one for summer (16 April until October 31) since 1994 (Figure 4.4). When ground-level ozone concentrations exceed or are expected to exceed 82 parts per billion (ppb), a warning is issued asking the public to reduce their use of motor vehicles, solvents, and volatile chemicals and to use public transit services. These warnings are issued 24 hours in advance as well as when actual ozone measurements indicate the presence of smog.

The Info-Smog program also makes daily air-quality forecasts based solely on ground-level ozone concentrations. There are three categories for air quality: 'good' is defined by concentrations less than 41 ppb; 'fair' by concentrations in the range of 41 to 81 ppb; and 'poor' by concentrations of 82 ppb and above. An Info-Smog warning is issued when the forecast category is identified as 'poor' (Environment Canada 2007d). Info-Smog warnings are available online, on a telephone hotline, and on television, but the program is also enriched by a partnership with the provincial ministry Transport Quebec. Info-Smog warnings are broadcast on the electronic panels overhanging highways in the Greater Montreal region as well as in the Quebec City area.

Geographical features that restrict air-flow patterns can contribute to the smog problems of an area. This is the case in the Lower Fraser Valley in British Columbia, and the Lower Fraser Valley Air Quality Monitoring Network, operated by the Greater Vancouver Regional District, has established a monitoring program in the region (Langley, Surrey, and Chilliwack). In 2003, smog levels in the Lower Fraser Valley

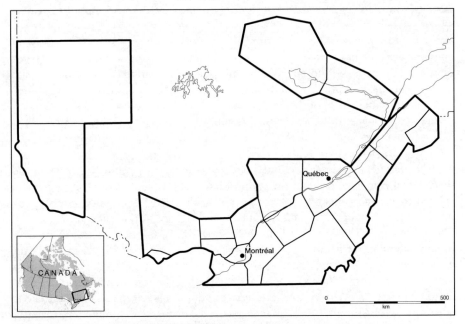

Figure 4.4 Regions of Quebec that are covered by the Info-Smog program.

Source: Modified from http://www.qc.ec.gc.ca/atmos/images/RegionInfoSmogAvril2002.gif.

exceeded safe levels (for human health) about 44 per cent of the time for ground-level ozone (Environment Canada 2007e).

Most of the other provinces tend to have good air quality. Alberta, for example, reports good air quality for its cities over 90 per cent of the time, according to the Air Quality Index (Clean Air Strategic Alliance 2007). Ground-level ozone in Manitoba is measured in urban areas like Winnipeg and Brandon, and while ozone concentrations at these sites are sometimes above the province's objective for air quality, the frequency of such occurrences is well below that encountered in Canada's problem areas. In fact, Manitoba has experienced a decrease in annual ozone concentrations over the last few years (Manitoba Conservation Air Quality Section 2007). Air pollution in Atlantic Canada is a function of both the long-range transport of air pollution and significant local emission sources. The same wind patterns and atmospheric conditions that bring the weather conditions from central North America and the eastern United States to Atlantic Canada also carry large concentrations of air pollutants to the region. Atlantic Canada is also highly dependent on fossil fuels such as coal and heavy oil, which are both significant emission sources, making the region a contributor to its own air pollution as well (Environment Canada 2007a).

Thunderstorms

Summer thunderstorms are fairly common in most areas of Canada, providing much-needed precipitation for gardens, crops, and other purposes. Unfortunately, a small percentage of them intensify to the extent that they become 'severe', damaging property and threatening lives. Severe thunderstorms in Canada are most frequent in Ontario, Quebec, and parts of the Canadian Prairies. Environment Canada issues a severe thunderstorm or tornado warning if heavy rain, high winds, tornadoes, hail, or lightning are present or expected. Intense lightning watches and warnings are issued in British Columbia, but not in any other province. The storm's expected motion and development are also provided in the weather warning (Environment Canada 2007j). Compared to other parts of the world, Canada fares quite well in terms of thunderstorm activity. Places like the American southeast, central Africa, and Indonesia can expect as many as 100 thunderstorm days per year, compared to an average of 5–10 for most parts of Canada.

Severe thunderstorms are most likely to occur when the atmosphere is unstable and there is abundant low-level moisture, when there is strong wind shear, and when there exists a trigger mechanism that can release instability. These conditions are strongly influenced by topography and air mass climatology (Etkin and Brun 2001). Thunderstorms fall into three main categories: air mass (single-cell) thunderstorms, multicell thunderstorms (or squall-line storms), and supercell thunderstorms. The *air mass thunderstorm* is a common and usually non-severe system that forms away from frontal systems. *Multicell storms* can form in a line known as a squall line, where continuous updrafts are fed by low-level convergence. *Supercell storms,* which include most of the severe thunderstorms, are characterized by massive cumulonimbus clouds that grow rapidly into individual thunderstorms. These supercells are sometimes embedded in a larger cluster of thunderstorms, particularly in eastern Canada. West of the Great Lakes, where there is less low-level moisture, supercell thunderstorms usually form in isolation (Bullock 2007).

A critical factor in the development of a thunderstorm is the *lapse rate* or the change in temperature of the ambient atmosphere with a rise in elevation. The greater the temperature difference, the more likely a thunderhead will expand upwards. Other factors in the development of thunderstorms are the amount of moisture available at low levels in the atmosphere and the presence of mechanisms of uplift (such as a topographic barrier or a powerful jet stream drawing air upwards). Computer models take all these factors into account, making forecasting more accurate than ever before (Burt 2004). While the latest high-resolution models have proved to be capable of simulating thunderstorms after the fact, the models used by forecasters do not explicitly forecast thunderstorms and are not particularly accurate in forecasting the location, character, or severity of a thunderstorm (Bullock 2007).

A research initiative now underway in Alberta intends to investigate boundary-layer processes associated with severe thunderstorm development over the Alberta foothills. The so-called Understanding Severe Thunderstorms and Alberta Boundary Layers Experiment (UNSTABLE) is a field study undertaken by researchers at the Cloud Physics and Severe Weather Research Section of Environment Canada and the Hydrometeorology and Arctic Lab of Environment Canada. The Canadian Prairies are subjected to a high frequency of severe thunderstorms, with an average of 203 severe weather reports received by Environment Canada each summer. The Alberta foothills are a preferential region for thunderstorm development, experiencing more thunderstorm days than any other region in the Prairie provinces. Most storms that develop there move eastward to affect the Edmonton–Calgary corridor. This is one of the most densely populated and fastest-growing regions in Canada, and consequently Alberta has become particularly susceptible to costly thunderstorm events; Public Safety Canada estimates that since 1981 more than 40 lives have been lost and $2.5 billion in damages incurred owing to severe thunderstorms (Taylor et al. 2007). It is hoped that UNSTABLE will, among other things, help us better understand the processes leading to the development of such severe thunderstorms. The results of the study will be shared with operational forecasters to help them improve accuracy and lead time for severe thunderstorm watches and warnings.

Hail

In August 2007, a destructive hailstorm hit Dauphin, Manitoba. The storm only lasted about 30 minutes but it caused significant damage. Over 13,000 insurance claims representing estimated losses of $53 million were made to Manitoba Public Insurance (MPI), making the hailstorm the most expensive single event that the MPI had ever had to deal with. This was only one of the 279 so-called hailers that affected the Prairies in 2007 (Environment Canada 2008a). In that year in Saskatchewan, the frequency of storms was up and so was the severity. In certain places, crops were totally destroyed, and there was extensive damage to homes, vehicles, and farm equipment. It is not surprising then that Saskatchewan producers now spend about $100 million to insure their crops against hail every year.

Hail is one of the most dramatic phenomena associated with severe thunderstorms. With respect to insured costs, hail ranks as one of the most costly natural hazards in Canada. On average, hailstorms destroy roughly 3 per cent of Canada's prairie crop each year. The regions in Canada that experience the most frequent hail are the central and eastern Prairies (parts of Alberta, Saskatchewan, and Manitoba), south-central British Columbia, and southwestern Ontario (Figure 4.5). This spatial distribution of hail places much of Canada's productive farmland and much of its population at risk (Etkin and Brun 2001). Hailstorms occur somewhere in Alberta an average of seven to 11 days per summer (Charleton et al. 1995; Smith et al. 1998). On 12 August 2006, a powerful thunderstorm swept through central Alberta, bringing severe hail that punched holes in the siding of buildings and broke windows. Just about every one of the 400 homes in Springbrook, Alberta, had damage of some sort, with either windows or vinyl siding completely ripped away. Drifts of slush from hail were piled up along highways, and many trees were stripped of their leaves. The damage was in the millions. The worst hailstorm in Canadian history in terms of paid insurance claims was the Calgary hailstorm of 7 September 1991. A 30-minute downpour resulted in 62,000 claims for property damage (largely broken windows), totalling $237 million, and a further 54,000 claims for vehicle damage (primarily dented roofs), totalling $105 million (PSC 2007a).

Hail consists of ice pellets that are formed in roughly concentric layers in the strong updrafts critical to the formation of thunderstorms. Hailstones are at least half a

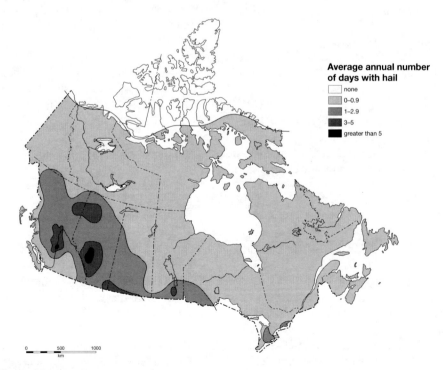

Figure 4.5 Average annual number of days with hail in Canada.
Source: Modified from PSC 2007a.

centimetre in diameter; below that measurement, frozen precipitation is defined as snow or ice pellets. Initially, an updraft carries water droplets above the freezing level to form the core of a hailstone. When the pellet falls from the updraft and collides with liquid droplets, the core is coated with a film of water. If the pellet is lifted by the updraft again, the process is repeated, enabling the hailstone to grow in diameter. The final size of the hailstone depends on the intensity of the updraft, with stronger vertical motions being capable of lifting larger hailstones (Aguado and Burt 2007). With very strong updrafts, hailstones can grow larger than 10 cm (about the size of a grapefruit). Thunderstorm activity tends to occur in warm weather, and therefore damaging hailstorms generally only happen in Canada from May to October (PSC 2005b).

Hail can hit the ground at 130 km/hr, causing severe damage to crops, houses, and vehicles, as well as injuries to people and animals (Figure 4.6). For many Canadians, a hailstorm is an intriguing rarity, but for farmers whose crops are crushed—or for those whose homes and cars are damaged—a hailstorm can translate into a financial burden (PSC 2007a). In order to address this, the so-called Hail Suppression Project was initiated in Alberta in 1996. (It should be noted, however, that the federal and provincial governments have determined to closely monitor and limit the use of weather modification until it is shown conclusively that such modifications do not in any way alter weather patterns and precipitation levels in non-target areas.) The project is intended to lessen the otherwise damaging impacts of hail events by seeding clouds with silver iodide particles via aircraft. The particles provide more surfaces upon which ice can accumulate, thus reducing the size of hailstones.

Figure 4.6 Photo of hail damage.
Source: Randy Fiedler/Red Deer Advocate.

There remains considerable controversy over such weather modification practices, however. If large-scale weather control were to become feasible, potential risks include the following:

- Unintended side effects, especially given the chaotic nature of weather development
- Damage to existing ecosystems
- Health risks to humans
- Equipment malfunction or accidents
- Non-democratic weather control or use of control as a weapon

Hail crop insurance dates back to 1905, and attempts by the agricultural community to suppress hail by cloud seeding dates back to the 1950s. In spite of opposition from some quarters, governments and insurance companies have sponsored cloud seeding to reduce the damage caused by hail.

Tornadoes

At the end of June 2007, a series of tornadoes battered southern Manitoba. Although more than eight large tornadoes touched down, only a few people were injured. This good fortune was attributed to the fact that people had known what to do and had quickly sought shelter in their basements. One of the tornadoes in the outbreak, however, levelled four houses and demolished a farm (*Chronicle-Herald* 2007b). This severe weather system prompted a call for a national tornado-warning system and also highlighted the fact that it was only a matter of time before Winnipeg would be struck by a similar storm. The first confirmed F5 tornado in Canada struck the small community of Elie, Manitoba, on 22 June 2007, with winds in excess of 400 km/hr (Figure 4.7). (The grading of tornadoes, from F0 to F5, is discussed below.) After reviewing amateur videotape of the tornado, scientists at Environment Canada confirmed its status as an F5 event. In the video, a house could be seen being picked up and thrown a few hundred metres through the air and then exploding. Fortunately, no one was seriously injured or killed. Local residents attributed this to the fact that the tornado struck during dinnertime, so many people were at home and could quickly take cover in their basements (CTV 2007). F5 tornadoes account for less than 1 per cent of all tornadoes in the world. Canada ranks second in the world for tornado occurrences, while the United States is number one with an average of 1000 to 1200 each year. While Canada had never recorded an F5 tornado before the event at Elie, the United States usually reports one or two yearly (Environment Canada 2008a).

The first reported tornado in Canada occurred on 1 July 1792 in what is now southern Ontario. Early settlers took advantage of the straight line of levelled trees and built a road along the tornado's path, which they interestingly named Hurricane Road. The first tornado warning was issued on 14 July 1950 by the Forecast Office in Regina, Saskatchewan. The citing of a tornado had been reported to the office by the crew of a TransCanada flight west of Johnstone Lake, Saskatchewan (Etkin et al. 2001). On Monday, 20 May 1996, a tornado ripped through the grounds of a drive-in theatre near Niagara Falls, Ontario, and ironically, the movie *Twister*, in which a tornado destroys a drive-in theatre, was on the bill!

Figure 4.7 Photo of tornado impact/damage at Elie.
Source: The Canadian Press/Wayne Hanna.

The province of Ontario has the greatest number of tornadoes, followed by Saskatchewan, Alberta, and Manitoba (Figure 4.8). Etkin et al. (2001) have determined that the spatial pattern of tornado events in Canada correlates well with population density, with the exception of the Maritime provinces. They also address the effects of El Niño–Southern Oscillation (ENSO) events on tornadic activity, and they highlight the need to establish a robust relationship between ENSO and tornadoes in Canada.

Newark (1984) was the first researcher to gather a data set of tornado characteristics and their climatology in Canada (using a 30-year period from 1950 to 1979). Most tornadoes in Canada occur between May and September, with peak frequency in July. An average of 80–100 tornadoes are reported each year, although the numbers reported vary considerably from year to year (Etkin et al. 2002; Environment Canada 2005d). Canada has essentially two 'tornado alleys'. One extends east of the Great Lakes from southern Ontario through southwestern Quebec and western New Brunswick. The thunderstorms that form there have a distinctly different character (owing largely to the greater available of moisture) from those that form in the other tornado alley, which runs from Alberta through Saskatchewan and Manitoba (Bullock 2007). The clustering of tornado observations in southern Ontario and Quebec and southern portions of the Prairie provinces has both climatological and demographical explanations, as tornadoes in unpopulated areas often go unreported (Etkin 1995).

According to Etkin et al. (2001), the actual mechanisms for tornado formation are localized, while the broader climatology of severe conditions is associated with prevalent storm formation mechanisms across Canada. In Saskatchewan, moist air coming from the south originates in the Gulf of Mexico. It flows over the Canadian Rocky Mountains and creates conditions that are conducive to supercell thunderstorm

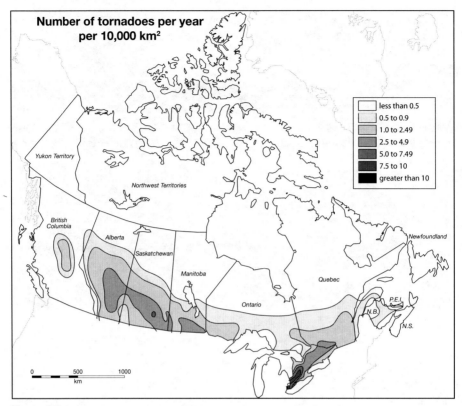

Figure 4.8 Distribution and number of tornadoes across Canada per year.
Source: Modified from Newark 1984.

formation (severe thunderstorms with a deep, continuously rotating updraft). The westerly wind flow also produces the high wind shear conditions conducive to the formation of severe weather. As the mid-latitude jet stream weakens and retreats during late spring in most years, moderate to occasionally strong westerly winds are sustained across Canada. Beneath this, the warm and humid air originating from the south provides low-level moisture (the 'fuel' for thunderstorms). The southerly winds bearing the low-level moisture, combined with the mid- and upper-level westerlies associated with the retreating and weakening jet stream, yield a highly sheared and buoyant atmosphere—one supportive of supercell thunderstorms and subsequently potential tornadic development.

Etkin et al. (2001) summarize several previous studies of tornadoes across Canada. The first, which examined tornadoes in western Canada between 1890 and 1958, indicated that tornadoes struck southeastern Saskatchewan and southwestern Manitoba more often than other parts of the Prairies, with 76 per cent of events occurring in June and July. Studies of tornadoes in Alberta and Saskatchewan between 1879 and 1984 (Hage 1987, 1994) indicated that there had been over 700 tornadic events and over 1000 non-tornadic severe windstorms in this period; Etkin et al. note that 90 per cent of the tornadoes occurred between 1 June and 15 August. Paul (1995)

Lightning strikes the CN Tower in Toronto.
Source: Sam Javanrouh. www.ddoi.ca

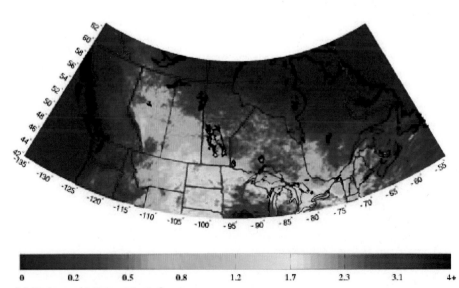

Satellite image of lightning strikes in Canada.
Source: Modified from Environment Canada 2003. © Her Majesty the Queen in Right of Canada, 2003. Reproduced with the permission of the Minister of Public Works and Government Services Canada.

Storm surge.

Source: The Canadian Press/St. John's Telegram/Keith Grosse.

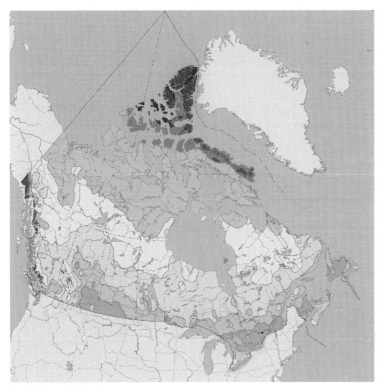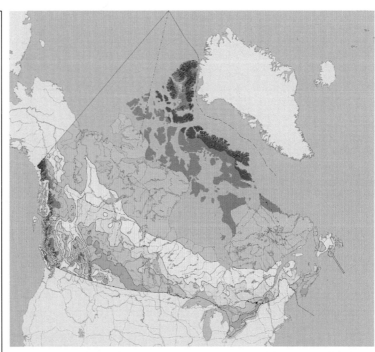

Median start date of onset of continuous snow cover (above) and median end date of continuous snow cover (below).

Source: Legends from nrcan and modified from http://atlas.nrcan.gc.ca.

Saguenay Flood.
Source: The Canadian Press/Jacques Boissinot.

A geostationary satellite image of Hurricane Bob (1991).
Source: NOAA/Science Photo Library.

The Gulf Stream.
Source: Reprinted by permission of NASA.

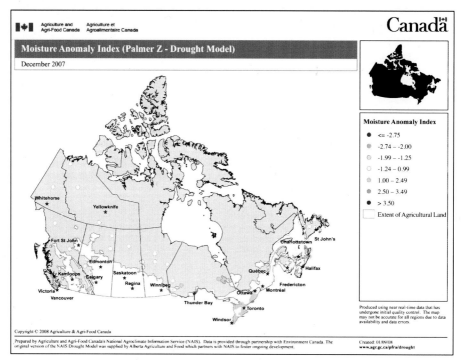

Moisture Anomaly Index (Palmer Z-Drought Model), July 2008.

Source: Moisture Anomaly Index (Palmer Z-Drought Model)—July 2008, Agriculture and Agri-Food Canada.

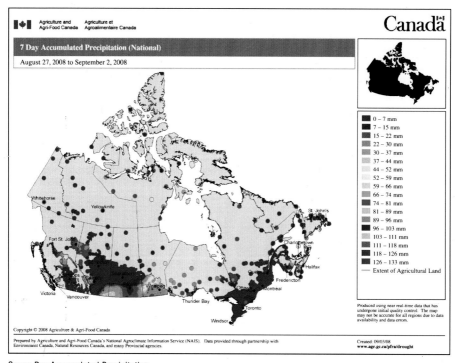

Seven Day Accumulated Precipitation.

Source: Seven Day Accumulated Precipitation (National) August 22, 2008 to August 28, 2008. Agriculture and Agri-Food Canada (2008). Reproduced with the permission of Agriculture and Agri-Food Canada, 2008.

A beautiful chinook arch.
Source: Photo by Rafal K. Komierowski.

The Golden Gate Bridge, blanketed in fog.
Source: Frederic Larson/San Francisco Chronicle/Corbis.

Garden City nocturne.
Source: Copyright © 2005 David Sills.

examines the 118 tornadoes reported in Saskatchewan from 1950 to 1990 (48 of them classified as strong). He estimates that up to 13 tornadic events occurred in the city of Regina, although an 'extremely conservative' estimate would be three events.

The majority of 'twisters' do little more than bend TV antennae, break windows, uproot trees, and damage weak structures such as barns and sheds. Tornadoes are categorized on the basis of intensity, a function of wind speed (Table 4.2). The more violent tornadoes are the most devastating storms on Earth. With winds approaching 500 km/hr, they can level even the most solid structures. The Fujita Scale has traditionally been used to rate the strength of a tornado on the basis of the damage it inflicts on property: F0 is the least intense, F5 the most intense. The scale was named for Dr T. Fujita, a pioneer in tornado research at the University of Chicago (Environment Canada 2005d). The National Oceanic and Atmospheric Association (NOAA) in the United States no longer uses the Fujita Scale and instead has moved to the 'Enhanced Fujita', which requires an investigator to assess the effects on several different types of structures (28 possibilities) before assigning a value.

Tornadoes are zones of extremely rapid rotating winds beneath the base of cumulonimbus clouds. Strong tornadic winds result from extraordinarily large differences in atmospheric pressure over short distances. Tornadoes can develop under any condition that produces severe weather frontal boundaries, squall lines, supercells, and

Table 4.2   The Fujita Scale.
Source: Modified from NOAA 2007d.

Scale	Wind Estimate (mph)	Typical Damage
F0	<73	*Light damage:* Some damage to chimneys; branches broken off trees; shallow-rooted trees pushed over; sign boards damaged
F1	73–112	*Moderate damage:* Peels surface off roofs; mobile homes pushed off foundations or overturned; moving autos blown off roads
F2	113–157	*Considerable damage:* Roofs torn off frame houses; mobile homes demolished; boxcars overturned; large trees snapped or uprooted; light-object missiles generated; cars lifted off ground
F3	158–206	*Severe damage:* Roofs and some walls torn off well-constructed houses; trains overturned; most trees in forest uprooted; heavy cars lifted off the ground and thrown
F4	207–260	*Devastating damage:* Well-constructed houses levelled; structures with weak foundations blown some distance; cars thrown and large missiles generated
F5	261–318	*Incredible damage:* Strong frame houses levelled off foundations and swept away; automobile-sized missiles fly through the air in excess of 100 m (109 yds); trees debarked; occurrence of incredible phenomena

tropical cyclones (Aguado and Burt 2007). Their precursors are generally warm humid weather and thunderstorms that develop when cool northern air masses collide with hot air flowing north from the Gulf of Mexico. When complex patterns of updrafts and downdrafts in the atmosphere are added, part of the base of the thunder cloud begins to rotate and a tornado is born (Environment Canada 2005d). *Downbursts* or 'plow winds' are also associated with severe thunderstorms. These are intense concentrations of sinking air that fan out when they strike the Earth's surface, producing straight winds with speeds up to 200 km/hr. Like tornadoes, downbursts can cause extensive damage. In fact, damage attributed to tornadoes has often actually been caused by a downburst. Such winds have also been known to produce a loud roaring noise, similar to that of a tornado.

Experts are not certain they can predict the conditions that cause rotation. Fortunately, rotation in a supercell usually precedes tornado formation by at least several minutes, which allows forecasters to issue tornado warnings with some lead time (Bullock 2007). To better understand the dynamics, researchers at Environment Canada study images of actual tornado clouds from radar, satellites, and photographs. This research will lead to more accurate forecasts and more timely warnings. Tornado warnings are important in reducing society's vulnerability. Unfortunately, providing timely tornado warnings to the population at risk is a challenging task. In the 31 May 1985 tornadic event in Barrie, Ontario, 91 per cent of affected people only had a warning of one minute or less (CBC Archives1985). Because of power outages caused by the tornado, radio and television were of no use, and the people who needed to be aware of the approaching tornado did not hear the weather warning issued by Environment Canada. In the Aylmer tornado of 4 August 1994, only 3 per cent of those affected heard the Environment Canada weather warning, and 74 per cent of the residents had no knowledge of the appropriate response to a tornado. Studies by White et al. (1995) and Carter et al. (1989) suggest that tornado warnings are not effective in terms of optimizing people's response to tornadoes, in Canada at least (Etkin et al. 2001). Much more should be done in this regard. In the United States, the network of public alert sirens installed during the Cold War to warn of impending nuclear attack is sometimes used to alert the public of an approaching tornado.

Not to be confused with tornadoes, *gustnadoes* typically appear as a swirl of dust or debris along the 'gust front' of a thunderstorm. They are not directly linked with the rotation in the thunderstorm itself and can form a considerable distance away from the parent storm. There is no condensation funnel or other visible connection to the cloud base. Gustnadoes account for a large number of the weakest tornado reports each year. Their localized impact and damaging effects have allowed them to be counted as tornadoes, but most are probably not 'true' tornadoes (Environment Canada 2005d).

Table 4.3 provides information on some of the worst tornado and hail events in Canada from 1879 to 2004. The tornado that struck Regina in 1912 remains on record as one of the most damaging in Canadian history. On 30 June 1912, many of the residents of Regina were preparing for the July 1st celebrations when a deadly tornado swept through, killing 28 people and leaving another 2500 homeless (CBC Archives 1976). The extensive damage to Regina's predominantly wooden structures suggests that the tornado was likely an F4 on the Fujita Scale. The nature of the structures

Table 4.3 Some of the worst tornado and hail events that have struck across Canada from 1879 to 2004

Source: Compiled and modified from CBC 2006b, 2006k, 2007a; PSC 2006a; Environment Canada 2005; Etkin et al. 2001.

Location	Date	Conditions
Buctouche, NB	6 Aug. 1879	Five to 7 dead and 10 injured; damage of $100,000; 25 families left homeless
Valleyfield, QC	16 Aug. 1888	Nine (possibly 11) dead and 14 injured; extensive property damage
St-Rose, QC	14 June 1892	Six dead and 26 injured
Regina, SK	30 June 1912	Known locally as the 'Regina Cyclone', this tornado is one of the worst in Canadian history in terms of death toll; 28 people die, and hundreds are injured as the tornado covers 30 km. Property damage totals $4 million, and 500 buildings are damaged.
Southeastern Saskatchewan	22 July 1920	Four people die and more than a dozen are injured when a tornado sweeps across the region.
Portage La Prairie, MB	22 June 1922	A tornado results in 5 dead, scores injured, and damage of $2 million.
Lebret and Kamsack, SK	1 July and 9 Aug. 1944	Two devastating tornadoes hit in the same summer, killing 4 people at Lebret and 3 at Kamsack. Almost 75% of Kamsack's homes and 100 businesses are destroyed.
Windsor, ON	17 June 1946	A tornado results in 17 dead and hundreds injured. More than 1000 buildings are damaged (damage of $100 million).
Central Alberta	14 July 1953	Thousands of birds are crushed by golf ball–sized hail as a storm moves across an 1800 km^2 area.
Sarnia, ON	21 May 1953	Seven people die and 40 are injured in a tornado that also results in over $10 million in damage.
Liverpool, NS	30 Jan. 1954	A rare winter tornado producing hail and lightning hits the coast of Nova Scotia near White Point Beach.
Lambeth, ON	19 Aug. 1968	A severe hailstorm causes extensive crop and property damage and leaves ice up to 17.5 cm thick on streets.
Sudbury, ON	20 Aug. 1970	An F3 tornado (wind speeds of 252–330 km/hr) takes 6 lives and injures 200. Damage is in excess of $45 million.
Western Prairies	23 July 1971	Heavy hail damage extends over 500 km, resulting in over $99 million in damage.
Southeastern, SK	11 Aug. 1972	Severe thunderstorms produce high winds and golf ball–sized hail. Yorkton has damage of $2 million.
Cedoux, SK	27 Aug. 1973	The largest documented hailstone in Canada is produced by this storm—290 grams and 114 mm in diameter. The storm causes more that $38 million in damage.

Table 4.3 Continued

Location	Date	Conditions
Windsor, ON	3 April 1974	Nine dead and 30 injured; damage of $500 thousand
St Bonaventure region of Trois-Rivières, QC	25 July 1975	A tornado that starts around 5:40 p.m. causes electrical failures and much damage; 75 dwellings are hit, leaving 45 uninhabitable; 918 people have to be relocated; 59 people are injured, and 3 people die.
Regina, SK	21 May 1979	A tornado tears off part of the roof of a wheat elevator. The funnel cloud also destroys a farmhouse and barn and sends a truck flying across the yard.
Montreal, QC	14 June 1982	Five people are killed and 26 injured when a tornado rips through the Ste-Rose region of Montreal, flattening hundreds of barns and homes.
Barrie, ON	31 May 1985	A state of emergency is declared and emergency centres are set up when 600 are left homeless. Eight people die and 155 are injured. Fortunately, children are let out of school early as the power has gone off; minutes later, the tornado rips the school apart.
Edmonton, AB	31 July 1987	A tornado touches down in eastern Edmonton, remaining on the ground for an hour and reaching wind speeds of 420 km/hr; 27 people are killed and property damage exceeds $330 million.
Southeastern Saskatchewan	8 July 1989	Tornado-force winds and hail cause considerable damage to buildings and cars in several towns and farms. In Peebles, the general store and the skating/curling rink are blown into the bush about 3 km away.
Calgary, AB	7 Sept. 1991	A supper-hour storm lasting 30 minutes drops 10 cm–diameter hail in Calgary subdivisions. Trees are split, windows and siding broken, and birds crushed. Homeowners file a record 116,000 insurance claims, with property damage exceeding $300 million. This is the most destructive hailstorm to date and the second costliest storm in Canada.
Saskatoon, Maymont, and Osler, SK	4 July 1996	Nine tornadoes strike, one of which is an F3. Homes and property are damaged.
Calgary, AB	24 July 1996	Hail the size of oranges clogs storm sewers and causes massive flooding. Eight days earlier, children make "hailmen" and go toboganning in July!
Winnipeg, MB	16 July 1996	Fist-sized hailstones bombard Winnipeg, resulting in $300 million in property losses. More than half the losses are for automobiles, making the storm responsible for the worst single-disaster claim against the Manitoba Public Insurance Corp. in its 25-year history.

Table 4.3 Continued

Location	Date	Conditions
Winnipeg, MB	5 Sept. 1996	A thunderstorm accompanied by hail and vicious winds topples 19 three-ton hydro transmission towers and threatens a massive blackout. Manitoba Hydro is forced to buy power from the US while crews spent several days repairing lines.
Okanagan region of British Columbia	21 July 1997	A destructive hail and wind storm rips through the orchards of the Okanagan, destroying or damaging nearly 40% of crops.
Drummondville, QC	6 July 1999	A tornado goes through the city of Drummondville, ripping roofs from 20 houses and injuring a number of people. One child is killed as a tree collapses on his tent. In total, 60 homes are damaged and 200 people are forced to flee from their homes. At least 4000 people are without power for 2 days.
Pine Lake, AB	14 July 2000	Approximately 140 people are injured and 400 campsites destroyed at the Green Acres Campground as a tornado rips through; 12 people perish. Wind speeds reached 300 km/hr and hail as large as baseballs is reported.
Edmonton, AB	11 July 2004	Severe hail and rain batters parts of Edmonton, causing extensive flooding of basements and streets and forcing the evacuation of 30,000 people from the world's largest shopping mall, the West Edmonton Mall. The storm causes $74 million in damages.
Eastern Manitoba	5 Aug. 2006	Three tornadoes move across eastern Manitoba, killing one woman in the resort community of Gull Lake. The storm wrecks farms and cottages across a wide swath and injures at least 20 people. It produces winds of up to 252 km/hr.
Southern and eastern Ontario and Quebec	1–2 Aug. 2006	Two people are killed in Quebec as violent thunderstorms sweep through, with winds gusting to 100 km/hr. Trees and hydro poles are knocked down, leaving 450,000 people without power. A tornado touches down on Highway 401.

largely explains why the storm did so much damage. The winds easily ripped through wood fastened with nails, in many cases causing building to collapse completely. Loosened planks became efficient and deadly projectiles, causing further damage and injury (Regina Plains Museum 2006). In addition to accounts of devastation, many amusing tales were told in the aftermath of the tornado, including one about a man who had been canoeing and was carried over 300 m through the air to be deposited on the shore still holding his paddle (CBC Archives 1976).

On Friday, 31 July 1987, another devastating tornado ripped through eastern Edmonton. What has subsequently come to be referred to as the 'Black Friday Tornado' remained on the ground for an hour, cutting a swath 40 km long and up to a kilometre in width. It was estimated that the tornado reached a strength of F4 on the Fujita Scale, killing 27 people and injuring more than 300. Property damage at four sites exceeded $330 million, and 300 homes were completely destroyed. It remains one of the worst natural disasters in Canadian history.

Advancements in computer technology since the 1960s have significantly improved meteorologists' ability to provide greater lead times in tornado forecasting. Doppler radar, in particular, has enabled meteorologists not only to detect areas of precipitation, but also to detect the wind circulations that may develop prior to a storm that produces a tornado. Today, severe weather forecasters use a combination of Doppler radar, enhanced satellite imagery, and sophisticated analysis programs to make essential life-saving decisions rapidly. Forecasters and researchers have at their disposal, in fractions of a second, a varying array of data that just 40 to 50 years ago would have been hand-plotted (NOAA 2007b).

Lightning

Southern Alberta is one of the most lightning-prone areas in Canada, receiving over half a million strikes per year. On 19 July 1996, lightning struck a truck near Alhambra, Alberta, and although the driver only felt a small jolt, the tires of his truck went flat over the subsequent four days, as the lightning had burnt out their steel belting (Environment Canada 2005d). Only two days earlier, a Calgary teenager had suffered minor burns when a lightning bolt tore a hole through the ceiling in her bedroom and set her bed on fire (Environment Canada 2005d). On 1 August 2006, lightning struck a retirement home in downtown Gatineau, Quebec, causing a major fire that left 85 elderly residents safe but homeless (CBC 2006f).

Lightning can be defined as 'an atmospheric discharge of electricity, which typically occurs during thunderstorms and sometimes during volcanic eruptions or dust storms' (NOAA 2007f). When charges reach sufficient strength to overcome the insulating threshold of the local atmosphere, lightning may occur (see photos in colour insert section). In thunderstorms, this process results in the accumulation of positive charges towards the top of clouds and an accumulation of negative charges in the base of clouds. The accumulated electrical potential is then neutralized by the electrical discharge either from one cloud to another or from a cloud to the ground. A flash of lightning occurs when a 'leader' from the cloud base meets an upward 'streamer' coming from the surface (Mills et al. 2008). A lightning bolt can reach up to 60,000 m/s in speed and 30,000°C!

Since February 1998, the Canadian Lightning Detection Network (CLDN) has provided continuous lightning detection in space and time, its range being to about 65°N in the west, to 55°N in the east, and to about 300 km offshore (Burrows et al. 2002). The CLDN consists of approximately 82 sensors that detect lightning over most of Canada, providing Canadians with the basic systems to detect and monitor this deadly facet of severe weather. Environment Canada and academic partners have recently initiated research that is intended to improve the understanding of the impact of lightning on Canadians in terms of health, property damage, service interruptions, and associated economic impacts (Mills et al. 2008).

Areas across Canada can be compared for their lightning activity on the basis of flash density (number of lightning flashes per year per km^2). Windsor, Ontario, receives approximately 251 flashes per km^2, whereas Halifax, Nova Scotia, receives only 28 flashes per km^2. Inuvik, Northwest Territories, receives the fewest flashes, with a flash density of only 1 flash per km^2 (Environment Canada 2003c). Lightning is most frequent in southern Ontario, southern Saskatchewan and Alberta, and areas off the southern coast of Nova Scotia (Burrows et al. 2002; Mills et al. 2008). In northern Canada and most of British Columbia, lightning is less common. When lightning occurs in Canada, it usually strikes in the warmest months of the year (May–October) and during the daytime. Interestingly, greater frequency of lightning days and higher flash densities coincide with more populous regions of the country (Mills et al. 2008).

Between 1991 and 1995, twenty-seven deaths were attributed to lightning; 11 occurred in Ontario, six in Quebec, six in the Prairie provinces, three in the Maritimes, and one in British Columbia (Bains and Hoey 1998). The majority of those killed by lightning strikes were men between the ages of 15 and 50 or individuals involved in outdoor activities (e.g., boating, golfing, hiking). Mills et al. (2008) expanded their study of lightning-related fatalities in Canada to cover 1921–2003. They found that a total of 999 lightning fatalities had occurred in Canada within this period, but that there had been a steady decline in mortality rates. Over 90 per cent of the deaths occurred in Ontario, Quebec, and the three Prairie provinces, with an average of 9–10 lightning-related deaths and 92–164 injuries each year in Canada. Lightning strikes can have other consequences, such as forest fires. Between 1986 and 2001, over 12,000 fires were ignited by lightning strikes in Canada (Mills et al. 2008).

Climate Change and Severe Summer Weather

With temperatures an average of 1.4°C above normal, the summer of 2006 was the second warmest Canada had experienced since nationwide records began to be kept in 1948 (Environment Canada 2006e). The eastern edge of the Northwest Territories and the western edge of Nunavut experienced the most startling temperature anomalies, with average temperatures more than 3°C above normal. Annual mean temperature has risen an average of 0.9°C in southern Canada over the last century. Associated with this increase in mean temperature is a relatively smaller increase in daily maximum temperature and a relatively larger increase in daily minimum temperature. Some researchers report that southern Canada has become not hotter, but less cold (Zhang et al. 2000). Bonsal et al. (2001) observe that while there have been increases in winter temperatures and increases in summer minimum temperatures, the trends in daily maximum temperatures have displayed little change (with only a few significant increases over the southern Prairies and northern regions). They also note that trends in Canada towards decreased day-to-day temperature variability are similar to global trends in this respect. Other researchers (e.g., Smoyer-Tomic et al. 2003) report that the frequency, duration, and intensity of heat waves in Canada are likely to increase. Those regions of Canada with the highest population densities and the most productive agricultural zones are expected to suffer the greatest increased risk of adverse heat wave effects.

Vincent and Mekis (2006) have updated the work of Zhang et al. (2000) and Bonsal et al. (2001) and conclude that the number of cold events has significantly decreased, while the number of warm events has significantly increased at many locations. The *European Journal of Public Health*, among other public health journals, recently reported that heat waves are one of the main health risks associated with climate change (Diaz et al. 2006).

Future implications for the Arctic will be of particular concern, given the expectation of reduced sea ice cover and the melting of permafrost. Reports from communities in the Arctic have indicated that the weather has become increasingly variable and unpredictable and that extreme temperature excursions relative to monthly means for a particular year (e.g., extreme variations within a month) are increasing in frequency in Alaska and northern Canada (Walsh et al. 2005). The Intergovernmental Panel on Climate Change (IPCC 2007b) reports that warming in the Arctic has been double that for the globe from the nineteenth to twenty-first century and from the late 1960s to the present.

Canada might learn from the experience of other countries. Unusually large numbers of heat-related deaths were reported in France, Germany, and Italy in the summer of 2003, probably the hottest summer in Europe since AD 1500 at the latest. Stott et al. (2004) state that it is an ill-posed question whether the 2003 heat wave was caused (in a simple deterministic sense) by climate change, because almost any such weather event can occur by chance in an unmodified climate. However, it is possible to estimate the degree to which human activities might increase the risk of the occurrence of such a heat wave. In unmitigated-emissions scenarios, summers like 2003 are likely to be experienced more frequently in the future. Anthropogenic warming trends in Europe imply an increased probability of very hot summers. Consequently, the risk that Europe may experience mean summer temperatures as hot as those in 2003 has more than doubled, and with the likelihood of such events projected to increase a hundredfold over the next four decades, it is difficult to avoid the conclusion that potentially dangerous anthropogenic interference in the climate system is already underway (Stott et al. 2004). Models indicate that future heat waves in areas of Europe and North America will become more intense, more frequent, and longer lasting in the second half of the twenty-first century (Meehl and Tebaldi 2004; IPCC 2007b). Although it will always be important to consider natural variability in our climate system, it is important to note that extreme heat wave events are extremely rare in an unmodified climate but become less rare with climate change (Lines 2007).

During the past few years, numerous studies of trends in extreme temperature and precipitation have indicated that there has been a significant decrease in the number of days with cold temperatures, an increase in the number of days with extreme warm temperatures, and a detectable increase in the number of wet days in many parts of the world (Vincent and Mekis 2006). Several studies have suggested that relatively small changes in mean temperature could result in substantial changes in extreme temperatures. More detailed analyses regarding the relationships among changes in means, extremes, and variances of Canadian daily temperatures are required. Climate models show a large increase in the number of heat waves and extremely warm days as a result of global warming (Cubasch et al. 2001). They have indicated, for example, that an increase of 4°C in Toronto's average temperature would likely increase the risk of summer days with temperatures exceeding 30.5°C from 1 in 10 to almost 1 in 2. Meehl and Tebaldi (2004) conclude that areas already experiencing strong heat waves could experience even more intense heat waves in the future, while areas (e.g., southwest Canada) not currently experiencing strong heat waves could begin to experience increases in heat wave intensity and the residents could suffer more serious effects from these because they would not be as well adapted to these higher temperatures. Redner and Petersen (2006) have investigated whether the magnitude and frequency of record-breaking temperatures could be influenced by climate change, such as global warming. They conclude that trends in extreme temperatures and weather events are likely to be influenced by climate change and that this should be an area of active research. They could not distinguish between the effects of random fluctuations and the effects of long-term trends in the frequency of record-breaking temperatures (with 126 years of data).

Several researchers have investigated the links between climate change and air quality, particularly smog. As climate changes, many regions are likely to suffer a degradation in air quality as a result of the combined effects of higher temperatures, more persistent stagnation, and increased emissions of temperature-dependent VOCs from vegetation. High levels of surface ozone and particle concentration have been associated with health impacts involving the cardiovascular and respiratory systems, so the issue of climate change and smog warrants our close attention (Mickley 2007).

In terms of the kinds of impacts we are seeing today, Dolney and Sheridan (2006) report a citywide rise in ambulance calls on hot days in Toronto; this observation is in line with the results of previous research, which showed that there is additional strain on human well-being during oppressive weather. Jacinthe Seguin, manager of the Climate Change and Health Office of Health Canada, cites recent Canadian research as evidence of the health impacts of climate change, noting, among other things, that warmer temperatures and extreme rainfall accumulations have contributed to outbreaks of waterborne disease in Canada. The Canadian Health Assessment 2007 (Health Canada) seeks to provide a better understanding of the vulnerability of Canadians to climate change and the capacity of institutions to adapt to it. Tol (2002) estimates that an additional 350,000 people worldwide could die from heat-related cardiovascular and respiratory problems per 1°C increase in the global mean temperature. Some research has shown a general decline in heat susceptibility over recent decades (e.g., Davis et al. 2002), whereas other research has shown that, among some of the most significant heat waves, recent ones have shown a closer association with high mortality rates (e.g., Kunkel et al. 1999). In either case, with a growing elderly population and increased social isolation, there remain significant numbers of people who will be susceptible to the heat, and in a potentially warmer world, heat susceptibility could increase even further (Sheridan and Kalkstein 2004; Kalkstein and Greene 1997).

According to an IPCC report (2007b), hot days, hot nights, and heat waves have become more frequent. Canadians can expect higher maximum temperatures and more frequent heat waves and droughts in the summer months, but there is insufficient evidence to tell whether they should expect similar trends with tornadoes, hail, and lightning. It is difficult to know whether climate change will lead to increased frequencies of tornadic events, since climate change models do not take small-scale phenomena such as thunderstorms, tornadoes, hail, and lightning into account (McBean 2005). Data suggests that tornado frequency in western Canada increases with positive mean monthly temperature anomalies. The inference is that if Canada's climate warms, then a corresponding increase in tornado frequency might be exhibited (Etkin et al. 2002). McBean argues that despite our limited knowledge of potential changes in the tornado hazard for Canada, Canadian disaster management strategies should nevertheless—on the basis of risk analysis and the implementation of the precautionary principle—be altered to take into account a greater potential threat of tornado impacts in the future.

The Canadian Arctic is expected to face some of the most dramatic changes as a result of climate change. During the past several decades, the Arctic has warmed at an alarming rate, and it is projected that it will continue to warm significantly. Global climate model experiments have indicated that the Arctic is a region that will warm more rapidly than the global average (Winton 2006a; IPCC 2007b). This warming trend has had a devastating impact on arctic ecosystems, affecting sea ice, permafrost, forests, and tundra. Warming has contributed to increases in lake temperatures, the thawing of permafrost, increased stress on plant and animal populations, and the melting of glaciers and sea ice. Research has revealed that sea ice extent and cover have been in decline since the 1950s and that this trend is accelerating. The decline is most pronounced in the summer, and if it continues at the current rate of decrease, the Arctic will have lost all of its perennial sea ice in about a hundred years.

Chapter Summary

Canada is not immune from severe summer weather. The impact of severe heat outbreaks, for example, can be substantial. The Prairies, southern Ontario, and the St Lawrence River Valley region of Quebec experience the highest incidence of heat waves in the country. Canada's worst smog corridor extends from Windsor through to Montreal, although smog days do occur in many other regions of the country as well. Severe thunderstorms in Canada are most frequent in Ontario, Quebec, and parts of the Prairies. Hail is one of the most dramatic phenomena associated with severe thunderstorms, with the central and eastern Prairies (parts of Alberta, Saskatchewan, and Manitoba), south-central British Columbia, and southwestern Ontario having the greatest number of hail days. Ontario reports more tornadoes than any other province, and is followed in this respect by Saskatchewan, Alberta, and Manitoba. Lightning is most frequent in southern Ontario, southern Saskatchewan and Alberta, and areas off the southern coast of Nova Scotia. According to recent climate change studies, Canadians can expect their summers to have higher maximum temperatures, heat waves, and droughts, although insufficient evidence exists to determine whether trends exist in tornadoes, hail, and lightning.

Chapter Five

Water: Too Much or Not Enough

Objectives

- To describe and account for the spatial distribution of heavy precipitation events across Canada
- To describe and account for the spatial distribution of flooding events across Canada
- To discuss some of the significant flood events that have occurred in Canada and to review flood management regimes
- To describe and account for the spatial distribution of fog conditions across Canada and to emphasize the impacts that severe fog events can have
- To describe and account for the spatial distribution of drought conditions across Canada and to review some of the severe droughts that have occurred
- To explain the procedures involved in monitoring drought and drought indices
- To discuss the implications of climate change for future precipitation, fog, and drought conditions in Canada

Introduction

By the end of June 2006, following an ideal early growing season of warm temperatures and just the right amount of precipitation and soil moisture, it looked as though the Canadian Prairies would have a bumper crop. But then, beginning in July, the Prairies were plagued with varying degrees of heat and drought, resulting in substantial crop-yield declines. This example underscores Canadians' reliance on suitable quantities of water for a healthy economy, environment, and society. Severe weather conditions (in the way of either too much or insufficient amounts of precipitation) can lead, in the worst case scenarios, to drought conditions or floods. Canada is by no means home to the world's wettest or driest conditions, however. Cherrapunji, India, achieved the record for the most precipitation in one year (and holds it still) when 26,470 mm of rain fell between August 1860 and July 1862 (Extreme Science 2008). This record makes Canada's wettest place seem relatively dry in comparison! Prince Rupert's average of 2468 mm of precipitation a year earns it the reputation of being Canada's wettest city (Environment Canada 2003e). Although many Canadians might think that the Atlantic coast receives large amounts of precipitation, east coast precipitation does not come close to the deluge on Canada's Pacific coast. One of the wettest locations on the Atlantic coast is Wreck Cove Brook with 1945 mm per year. Charlottetown receives 1173 mm of precipitation on average, and Sept-Isles (Quebec), Stratford (Ontario), and Cameron Falls (Alberta) are some of the wetter spots in other provinces, with 1156 mm, 1064 mm, and 1103 mm of annual precipitation respectively (Environment Canada 2006a).

On the other end of the spectrum, the driest place on Earth is the Atacama Desert in northern Chile, where it typically rains a few times a decade, receiving <1 mm/yr. According to NASA (2005), this is the only place on Earth that we know of where the Viking Lander could have landed, scooped up some soil, and failed to find evidence of life. Arctic Bay, Nunavut, is the driest place in Canada with only 12.7 mm of precipitation in 1949 (Natural Resources Canada 2004).

Heavy Precipitation Events

In terms of high totals of precipitation on individual days, a few places on the west coast have received more than 254 mm, more than some areas of Canada get in a year. British Columbia has the highest maximum daily precipitation value in all of Canada, with 489 mm in one day (near Ucluelet on Vancouver Island). The explanation for the high values on the west coast can sometimes be attributed to a phenomenon called the 'Pineapple Express', a Pacific Ocean jet stream that brings warm, moist air from Hawaii (hence the 'pineapple' reference) to the west coast of North America. The combined effect of the warm, moist air and orographic uplift along the mountain ranges of the coast causes some of Canada's most torrential rains to occur in the region. On 11 June 2006, the *Globe and Mail* reported 'Southern B.C. awash in Pineapple juice' as more rainfall fell than had ever been recorded before. Two hundred homes along the Chilliwack River were evacuated because of rising water levels. On 7 November 2006, the province's coastal counties received between 50 and 175 mm of rainfall. Flood warnings were in place as rivers discharged huge amounts of precipitation from their catchments. Such storms can pose a dual threat to coastal British Columbia: precipitation falls as rain instead of snow in the mountains, increasing the runoff into local rivers, and it also melts the snow lying at lower altitudes, adding even more water to swollen rivers. Table 5.1 provides information on some of the largest single-day rain events in Canada. All of the single-day rain events occurred in British Columbia. Six of Canada's 25 largest urban centres have an average of over 125 rainy days a year; they are all located in close proximity to a significant water source (Table 5.2).

The *Rainfall Frequency Atlas of Canada* (Hogg and Carr 1985) shows that all of the provinces and the southern regions of Yukon and the Northwest Territories have experienced one or more intense rainfall events (rainfall exceeding 25.4 mm). There is no standard definition for an intense rain event, in part because local rainfall thresholds vary considerably across Canada. Precipitation between April and October (the growing season) falls in the form of rain in most parts of Canada. There can be

Table 5.1 Some of the heaviest rain events recorded for a single day in Canada.

Source: Modified from Environment Canada 2006a. © Her Majesty the Queen in Right of Canada, 2006. Reproduced with the permission of the Minister of Public Works and Government Services Canada.

Location	mm	Date
Seymour Falls, BC	314	14 Jan. 1961
Seymour Falls, BC	300	10 Nov. 1990
Port Renfrew, BC	293	23 Feb. 1986
Hartley Bay, BC	276	8 Oct. 1991
Nitinat River Hatchery, BC	257	7 Nov. 1995
Seymour River Hatchery, BC	255	10 Nov. 1990

Table 5.2 Canadian cities that receive more than 125 rainy days in a year.

Source: Modified from Environment Canada 2006a. © Her Majesty the Queen in Right of Canada, 2006. Reproduced with the permission of the Minister of Public Works and Government Services Canada.

City	Days a Year with Rain
Abbotsford, BC	171
St John's, NL	162
Vancouver, BC	161
Victoria, BC	150
Halifax, NS	132
Sherbrooke, QC	128

considerable spatial variation in precipitation across the nation and over short distances as well. Figure 5.1 shows the general trends in precipitation across North America. A clear pattern of high amounts of precipitation can be seen across the Cordilleran region of British Columbia and Yukon, with the heaviest amounts on the western side of the mountain regions. In addition, the relatively dry Prairies and Arctic regions emerge in the spatial patterns. Moderate amounts of precipitation are found throughout Ontario, Quebec, and the Atlantic provinces. In addition to the spatial patterns, there are seasonal temporal precipitation trends across Canada that are not depicted on this map. The Pacific coast experiences a pronounced late fall–winter maximum, and the summers are quite dry in southern British Columbia. Across the Prairies, the early summer is a time of maximum precipitation, although annual precipitation is unpredictable from one year to the next. Most of the eastern regions of Canada do not experience a marked seasonality in their precipitation patterns (Bailey et al. 1997).

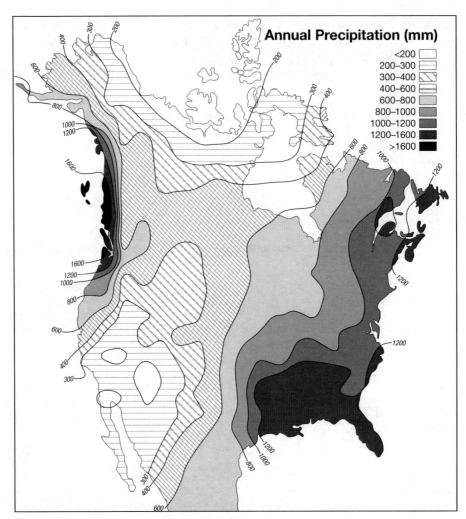

Figure 5.1 General trends in precipitation across North America.
Source: Modified from Groisman and Easterling 1994.107

Processes

All precipitation originates in parcels of moist air that rise to an altitude where the dew-point temperature is reached, air becomes saturated, condensation occurs, and clouds develop. At least one of four processes must take place in order for air to rise to the point where there would be significant precipitation: (1) the forced lifting of air where low-level winds converge; (2) the spontaneous rising of air, or convection; (3) the forced uplift of air as a result of topographic barriers; and (4) the forced uplift of air at the edges of colliding air masses (de Blij et al. 2005; Aguado and Burt 2007).

Convectional precipitation is a typical feature of the Canadian Prairies. It originates in convective clouds (i.e., cumulonimbus or cumulous), with precipitation initially falling as showers and then rapidly changing in intensity. During the growing season, precipitation on the Canadian Prairies is largely controlled by thunderstorm activity, which is often random and results in high variability in precipitation totals from one year to the next. In its study of precipitation on the Canadian Prairies from 1960 to 2004, Agriculture and Agri-Food Canada (2007) found that none of the years could be described as an extreme wet year (average precipitation being in excess of 200 per cent of the average for the entire study period). Raddatz and Hanesiak (2007) analysed nearly 1000 significant (>10 mm in 24 hours) summer rain events that occurred within the boundaries of the Canadian Prairie provinces from 2000 to 2004. The objective of this examination was to identify the mode of each event (i.e., solely or partially convective versus non-convective) and its primary forcing mechanism (i.e., source of lift). Daily rainfall and lightning maps reveal that most of the significant rain events (79 per cent) were solely or partially convective (i.e., lightning was recorded during the event).

Orographic precipitation accounts for the particularly heavy precipitation on the west coast of Canada. Figure 5.2 illustrates the rain shadow effect that results to the east of the BC coast. Secondary orographic precipitation occurs as air rises over the Rocky Mountains towards the province's eastern border with Alberta. Topography in particular affects rainfall in the Vancouver region, which sprawls across a flat river delta to the foot of a coastal mountain range. Precipitation data for Vancouver are measured at Vancouver International Airport, which is located on an island in the Fraser River. But not far from there, at the base of Grouse Mountain in North Vancouver, rain gauges fill much higher. The North Vancouver weather station averages 2477 mm of precipitation a year, twice the amount that falls at the Vancouver airport.

Frontal precipitation accounts for rain events in many regions across Canada. It occurs as a consequence of the slow ascent of air in synoptic systems, such as along cold fronts and in advance of warm fronts. When a warm air mass encounters a cooler one, the lighter warmer air overrides the cooler air. This produces a boundary called a warm front. Warm fronts, because of their gentle upward slope, are associated with light to moderate precipitation. In contrast, cold fronts are produced when cold air at the surface forces warmer air upwards, assuming a steeper slope and thus causing more abrupt cooling and condensation and triggering a narrower but more intense band of precipitation. Figure 5.3 illustrates the source regions of the predominant air masses that affect North America and in large part explain the annual precipitation patterns across the continent. The Prairies and the Arctic are dominated by dry continental air masses, while western and eastern regions receive moisture from maritime air masses. Typically, poleward-flowing tropical air masses collide with polar air masses moving towards the Equator. Upon convergence, the warmer air mass is forced to rise, creating a frontal boundary where clouds and precipitation will develop.

Floods

When viewed in the global context, floods in Canada tend to be less severe and hazardous than those in other parts of the world (e.g., monsoon-induced floods in southeast Asia) (Khandekar 2002). One of the worst disasters in recorded history occurred in August 1931, when the Huang He (Yellow) River in China flooded, killing an estimated

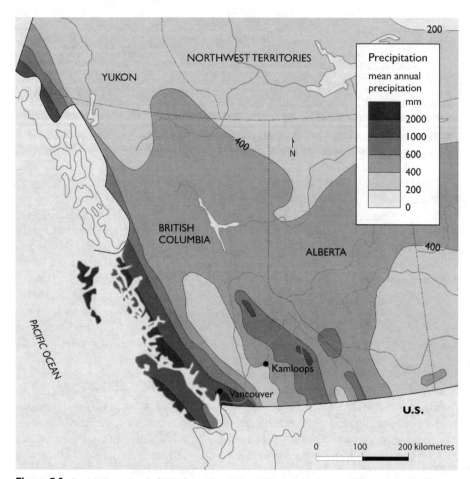

Figure 5.2 Precipitation pattern in British Columbia, with the distribution of isohyets exhibiting the results of the orographic effect as the westerly winds off the Pacific are forced across the north–south-trending Coastal-Cascade Mountains.
Source: Cathy Conrad.

3,700,000 people. Flooding in Canada has caused, directly or indirectly, the deaths of at least 198 people and can be counted among the country's most costly natural disasters, resulting in over several billion dollars of damage over the course of the twentieth century (Natural Resources Canada 2006). Between 1975 and 1999, sixty-three floods resulted in payouts of almost $720 million through the federal government's Disaster Recovery Financial Assistance Arrangements program (Shrubsole 2000).

Flooding can occur almost anywhere at any time of year, although some areas have been historically more vulnerable than others. According to Brooks et al. (2001), the 10 provinces and two territories (Nunavut is included within the Northwest Territories) experienced 168 known flood disasters between 1900 and June 1997 (Figure 5.4). In central and eastern Canada, flood disasters have primarily occurred in the south, where the population is concentrated. The distribution of disasters is much more scattered in western Canada, although there is some clustering in southwestern

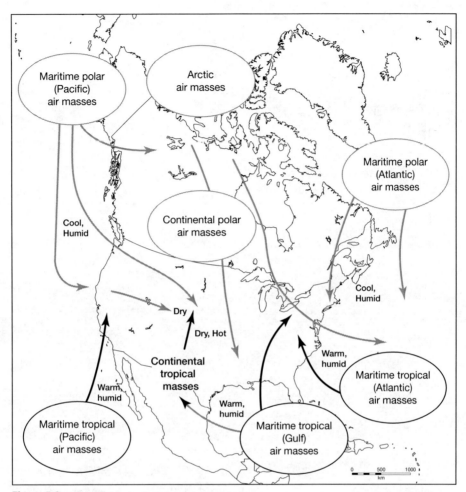

Figure 5.3 Source regions for North American air masses.

Source: Modified from Aguado, Edward and James Burt. *Understanding Weather and Climate*, 4th edn., © 2007, pp. 249, 257. Reprinted by permission of Pearson Education Inc., Upper Saddle River, NJ.

and northwestern British Columbia (Figure 5.5). Four provinces account for approximately 62 per cent of the disasters: Ontario (37 events), New Brunswick (26 events), Quebec (23 events), and Manitoba (18 events). One-quarter of Canada's most severe floods over the last two centuries have occurred on Manitoba's Red River floodplain. In 1996, the Saguenay region of Quebec experienced the most costly flood in Canadian history, suffering $1.5 billion worth of damages. The Fraser Valley of British Columbia is another floodplain at significant risk. In early June 2007, large areas around Smithers, Prince George, Terrace, and the Mount Currie reserve north of Whistler were under a flood watch because several days of intense heat had produced extremely high rates of snowmelt (CBC 2007b).

The province of Quebec has been particularly affected by flooding. With only 24 per cent of Canada's total population, it received 36 per cent of the federal-provincial-

112 Severe and Hazardous Weather in Canada

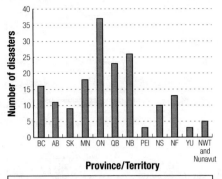
Fig. 1 Graph showing the occurence of known flood disasters in Canada for the period 1900 to June, 1997, by province (after Brooks et al. 2001, used with permission). The data are from the EPC Canadian disaster database, modified by Brooks et al. (2001).

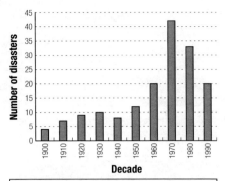
Fig. 1 Graph showing the occurence of known flood disasters in Canada for the period 1900 to June, 1997, by decade (after Brooks et al. 2001, used with permission). The data are from the EPC Canadian disaster database, modified by Brooks et al. (2001).

Figure 5.4 The distribution of floods by province and territory from 1900 to 1997.
Source: Modified from Brooks et al. 2001.

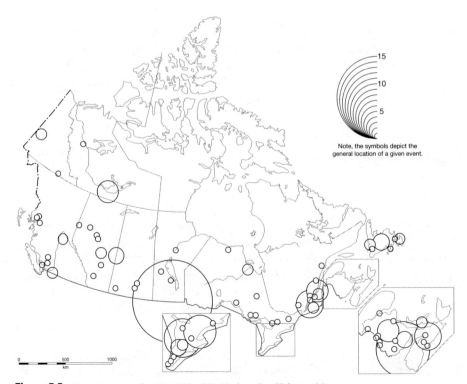

Figure 5.5 Flood disasters in Canada, 1900–1971 (single and multiple events).
Source: Reproduced with the permission of Natural Resources Canada 2008, courtesy of the Geological Survey of Canada.

territorial cost-shared disaster assistance arrangements for the period from 1970 to 1988, and roughly 51 per cent for the period from 1988 to 2003 (Roy et al. 2003). New Brunswick is also prone to flooding, particularly those communities and cities along the St John River. In early May 2008, over 1300 homes in the Fredericton area were threatened by floodwaters that rose 8.33 m above sea level. This event came dangerously close to the historic flood of 1973, when the river rose 8.61 m above sea level, causing $12 million in damage (*Chronicle-Herald* 2008). At the time of writing, the damage estimates and costs to the city are still be calculated but are expected to exceed those of the 1973 flood. Residents anticipate federal and provincial aid as well as assistance from relief organizations such as the Canadian Red Cross, since flood damage is not covered by insurance.

Causes and Examples of Flooding

The many causes of flooding in Canada include the following, alone or in combination: snowmelt runoff, rainfall, ice jams and other obstructions, catastrophic outbursts, coastal storms, and urban stormwater runoff.

Snowmelt Runoff

During the winter, most of the precipitation across Canada is stored as snow or ice on the ground. Snowmelt floods usually depend on these frozen stores, which have accumulated over a period of several months, and on the sudden occurrence of high temperatures (Kunkel et al. 1999a). During the spring melt, vast quantities of water are released in a short time, which explains the heavy spring runoff and flooding. Snowmelt runoff floods, the most common type of flooding in Canada, generally occur in the spring but also during sudden winter thaws. The water produced by the melting snow cannot penetrate the frozen ground and thus runs off over the ground surface into streams and lakes. If the snowmelt runoff is compounded by runoff from heavy rainfall, the situation can become a greater problem (a phenomenon that increases in likelihood during the late winter and early spring). Since the climatic factors influencing the rate of snowmelt are often widespread, snowmelt runoff flooding can occur over vast areas (Environment Canada 2004c). A database of flood disasters in Canada between 1900 and 1998 (Brooks et al. 2001) reveals that over 65 per cent of the flood disasters were the result of snowmelt runoff, storm rainfall, or 'rain-on-snow' (combinations of snowmelt runoff and storm rainfall).

Rainfall

The types of precipitation events that cause flooding are determined by the size and character of the river basin where the precipitation occurs. Single intense thunderstorms lasting only an hour or two are capable of producing flooding on very small basins, particularly those with high topographic relief, which promotes rapid runoff. These events can create flash floods, which characteristically peak within six hours of the onset of rainfall. Flood conditions develop rapidly when the rainfall is so heavy that the ground cannot absorb the water quickly enough, resulting in a very high runoff rate. The events are generally locally intense, and damage is usually restricted to a limited area. On moderate-sized basins, flooding is often caused by the occurrence of a series of moderate to heavy episodes over a few days. Flooding on the main stems of major rivers requires excessive precipitation over a period of several weeks to months (Kunkel 2003). The link between large amounts of precipitation and hydrologic flooding is complex and depends on several factors, including soil moisture, rate of melt in snowmelt floods, the character of the precipitation event(s) (intensity, duration), and the physical characteristics (size, topography, control structures) of drainage basins. These factors vary from event to event, season to season, and region to region (Kunkel et al. 1999a). Floods caused by intense rain events are typically local in nature; they occur frequently in coastal regions of British Columbia in the fall and winter seasons and in southern Ontario and Quebec during summer thunderstorms (Khandekar 2002).

The Saguenay flood of 1996 stands out as one of the most devastating floods in Canadian history, resulting in 10 deaths and the destruction or damage of 1718 houses and 900 cottages (see photo 'Saguenay flood' in colour insert section). It was Canada's first billion-dollar natural disaster. The flood occurred following an unprecedented heavy rainfall. Between the 19th and 21st of July, 200 mm of rain fell over a 5000 km^2 area in the Saguenay River basin. The two-day rainfall was equivalent to the volume of water that moves over Niagara Falls in four weeks (*Canadian Geographic* 2007). The large low-pressure system brought vast amounts of precipitation from the Caribbean towards Quebec's Saguenay River and Lac Saint-Jean. The storm produced an overland surge of water that transported rocks, trees, and mud; reservoirs were filled to capacity; and the river ripped away at cliffsides, bridges, and buildings. Residents were forced to evacuate their homes as water levels rose, and violent flows of mud-laden water poured into some of their basements. Many homes had been built on unstable land that city planners had known might be flood-prone but had nevertheless conceded to developers (Phillips 1997; Kunkel 2003).

The 1997 Red River flood in Manitoba stands out for its cost and extent. The Red River, 877 km long and located in the geographic centre of North America, is prone to frequent floods in part because in the spring its headwaters, located in the south, thaw before its northern Manitoba reaches are free from ice. Records indicate that several major floods occurred in the 1800s and 1900s. The earliest recorded flood in the Red River watershed was in 1826, although anecdotal evidence refers to larger floods in the 1700s (Simonovic and Carson 2003). The flood of 1997 was preceded by an autumn (1996) that had heavy precipitation (10 cm above average) and high soil moisture and a winter (1996–7) that was unusually severe, with a record snowfall. These factors created conditions conducive to major spring and summer flooding. Floodwaters covered about 5 per cent of Manitoba's farmland. In contrast to the 1950 flood, which had cost $606 million and had forced 100,000 people from their homes, the 1997 flood saw the evacuation of 28,000 Manitobans and flood damages exceeding $150 million. There remains a significant risk of flooding in the Red River watershed, and it has been indicated that the flood forecasting process needs to be improved and made more transparent to the public (Simonovic and Carson 2003).

Ice Jams and Other Obstructions
The floods caused by ice jams in rivers occur in the spring and are common in eastern Canada and northern Ontario. Ice jams are the result of the accumulation of ice

Figure 5.6 Image from Badger: town encased in ice.
Source: The Canadian Press/The St. John's Telegram/Keith Grosse.

fragments that first restrict the flow of water and then build up to act as temporary obstructions. Jams form during both the freeze-up and breakup periods, but it is usually the breakup jams that have the greater flooding potential.

Ice jams are a major cause of flooding in Canada. In the Saint John River basin in Atlantic Canada, for example, over two-thirds of total provincial flood damage costs are due to ice-related events (Environment Canada 2004c). Residents on the east coast will remember the flooding in Badger, Newfoundland, for its unique winter characteristics (Figure 5.6). More than 1000 residents were forced to evacuate from their homes when a massive ice jam sparked major flooding from three rivers, rapidly leaving their town encased in ice and water. Homes and businesses were destroyed, and contaminated water and sewer systems further necessitated the evacuation. The town of Badger experienced flooding in 1978, 1982, and 1985. The community is located on the floodplain of the Exploits River and has been identified by the provincial government as a 'flood risk zone' prone to flooding by ice jams and high spring runoff.

Catastrophic Outbursts

Outburst floods are common in western Canada, where lakes dammed by glaciers or moraines may suddenly drain, releasing tonnes of water, mud, and debris. An *outburst flood* typically occurs when the water level has become high enough to float the ice or when a small channel forms under the ice; the result is rapid melting and the sudden discharge of the accumulated water. The floodwaters can pick up large quantities of sediment and be transformed into destructive *debris flows*. At Kicking Horse Pass, British Columbia, in 1978, debris flows triggered by an outburst flood from Cathedral Glacier destroyed three levels of Canadian Pacific railway track, derailing a freight train. Sections of the Trans-Canada Highway were also buried (Environment Canada 2004b).

Coastal Storms

Hurricanes (or their remnants) are a significant flood mechanism in the Maritime provinces. They will be discussed in detail in Chapter 6. Many Canadians living along the shores of major lakes, such as the Great Lakes, or along Canada's coasts have experienced flooding and property damage as a result of the high wind and wave action associated with severe storms. Shoreline flooding may be caused by storm surges, or seiches, which often occur simultaneously with high waves (Environment Canada 2004a).

Urban Stormwater Runoff

Urban stormwater runoff can cause the flooding of local rivers as well as the flooding of the urban area itself. Stormwater runoff occurs when precipitation from rain or snowmelt flows over the ground. Impervious surfaces like driveways, sidewalks, and streets prevent stormwater from naturally soaking into the ground. Urbanization drastically alters the drainage characteristics of drainage areas by increasing the volume and rate of surface runoff. While the impact on major river systems may be minimal, the carrying capacity of small streams may be quickly exceeded, causing flooding and erosion problems. Often, the runoff from intense rainfall exceeds the carrying capacity of the sewer system, creating a backup in the system that results in the flooding of basements and roads (Environment Canada 2004e).

Although there can be flood disasters in any month of the year, about 40 per cent of them occur in April and May, the period that coincides with the snowmelt throughout much of southern Canada. This is also the period when several common flood mechanisms (snowmelt runoff, storm rainfall, and ice jams) are likely to occur, increasing the likelihood of high flows. Many of the flood disasters during the January to March period are the result of rain-on-snow during winter mild spells, while floods during the June to November period are the result of rainstorm runoff. Eleven of the 20 flood disasters in eastern Canada (Ontario to Newfoundland) in the months of August, September, and October were caused by hurricanes or their remnants. The smallest number of flood disasters have occurred in the months of November and December (Natural Resources Canada 2006).

The rapid expansion of Canada's population during the 1950s and 1960s gave rise to significant urban development, especially near bodies of water, considerably increasing the population's exposure and vulnerability to flooding (Robert et al. 2003). Table 5.3 includes some of the most notable floods in Canada's recent history.

Flood Management

In a recent study, Shrubsole (2007) describes flood management in Canada. There have been three phases: first, a structural control phase (from 1953 to 1970); second,

Table 5.3 Significant floods from the nineteenth and twentieth centuries.

Source: Compiled and modified from PSC 2007c; Canadian Red Cross, 2006.

Location	Date	Conditions
British Columbia	1894	The greatest Fraser River flood in the past century occurs when the floodplain is sparsely populated and undeveloped. Had the same flood struck the lower Fraser in 2000, it could have caused damages of $7.5 billion.
New Brunswick	1923	A flood is caused by snowmelt, heavy rain, ice, and log jams and is significant in all parts of the province. Two lives are lost.
British Columbia	May 1948	The Fraser River rises to within a foot of the 1894 level and floods more than 22,000 hectares.
Manitoba	April–May 1950	The 51-day Red River flood in and around Winnipeg is the result of snowmelt and heavy rain and causes major damage despite extensive diking. One person dies, and 107,000 are evacuated.
Ontario	Oct. 1954	Hurricane Hazel induces the worst flooding in the Toronto area in more than 200 years. The toll includes 81 dead and more than 20 bridges destroyed.
Nova Scotia	Jan. 1956	Snowmelt, rain, and ice jams cause extensive province-wide damage, including the destruction of more than 100 bridges.
Saint John River basin, NB	2–6 Feb. 1970	Snowmelt, ice jams, and heavy rainfall cause flooding in New Brunswick; a series of ice jams in six rivers destroys 32 bridges and damages 124 other bridges. Extensive damage is done to private homes, farmsteads, farm equipment, fishing gear, and industrial and commercial companies. Over 600 properties are damaged. One person dies when a barn collapses, and two others drown.
New Brunswick	April 1973	A frontal storm in the northern and central parts of the province unleashes the largest flood since records have been kept. Damage to Fredericton area and farmland accounts for about 60% ($77.8 million) of total.
Quebec	May–June 1974	Floods caused by an unusually wet spring and excessive snowmelt strike hundreds of towns, with the Ottawa River basin and Montreal region hardest hit. Total damages are $359 million. Over 1000 homes and 600 cottages are flooded and 10,000 people are evacuated.
Saskatchewan	April 1974	Runoff from a heavy snowpack floods the Qu'Appelle River basin and drives the Moose Jaw River to a historical high. In Moose Jaw, 60 city blocks are under water; bridges and dams are damaged.

Table 5.3 Continued

Location	Date	Conditions
Montreal	July 1987	A summer thunderstorm produces rainfall amounts of well over 100 mm within a few hours. Major expressways are flooded with over 3 m of water; in 30 minutes the subterranean Decarie Expressway is flooded with 3.6 m of water; the sewer system backs up, flooding approximately 40,000 homes and businesses; the subway system is closed; 2 people die, one from drowning in his car and the other from electrocution.
Lesser Slave Lake area, AB	5–8 July 1988	A severe rainstorm that started on 5 July brings up to 160 mm of rain to the town of Slave Lake and surrounding areas. This causes overbank flooding and log jams on Sawridge Creek, Driftpile River, and Swan River. Individuals, small businesses, farms, a hospital, and government property sustain damage. Major highways north of Edmonton to Slave Lake are closed. Over 800 homes are flooded, 500 severely damaged. Approximately 2800 residents are evacuated from the area.
Liard River, NT	1–9 May 1989	Flooding caused by severe ice jams near Three Mile Island damages the community of Fort Liard. A state of emergency is declared on 1 May owing to high water levels, 1 m in some areas. Water builds up when the ice jam reaches Fort Simpson, and the town of Hay River also suffers damage. Personal, municipal, and territorial property is damaged; riverbanks are eroded. Over 50 homes are affected by flooding, and 125 people are evacuated.
Winnipeg, MB	July–Aug. 1993	City of Winnipeg is declared a disaster area after prolonged heavy rainfall and 3 thunderstorms. Sewers back up, homes are damaged, power lines are down, and agricultural land and infrastructure are damaged. It is the wettest summer in 120 years.
Southern Alberta	June 1995	Heavy rain and meltwater lead to flooding of Oldman and Saskatchewan Rivers. Roads, property, riverbanks, agricultural land, and 20 bridges are damaged; 250 homes are flooded.
Saguenay River Valley, QC	July 1996	Ten people die and 15,825 are evacuated when floodwaters wash out thousands of homes, businesses, roads, and bridges. The flooding was caused by a sustained downpour of 290 mm of rain over 36 hours. Estimates of the total cost exceed $1.5 billion.
Manitoba's Red and Assiniboine River Valleys	May 1997	Thousands of volunteers, including residents, military personnel, and volunteers, work together for over a month to battle spring floodwaters and evacuate 25,000 people from the dozens of affected communities.

Table 5.3 Continued		
Location	**Date**	**Conditions**
Vanguard, SK	3 July 2000	330 mm of rain fall in 8 hours, causing extensive flooding. Roads leading into the community and surrounding area are either destroyed or made impassable. There is significant property damage both in the town and in the surrounding farmland.
Badger, NL	15 Feb. 2003	More than 1000 residents are forced to evacuate from their homes when a massive ice jam sparks major flooding from three rivers, leaving their town encased in ice and water.
Alberta	Jan. 2005	Heavy rainfall and high water cause a 'once in 200 years' flood event. The flooding damages hundreds of homes, as well as roads, bridges, and local infrastructure, in many parts of the province. More than 21 communities declare a local state of emergency.

a phase involving a reliance on a mix of structural and non-structural approaches (from 1970 to 1998); and third, a phase incorporating the proposed Natural Disaster Mitigation Strategy (the present period). An increasingly environmentally aware public, together with the realization that flood-control structures had not prevented significant floods, led to the shift in the early 1970s that was embodied in the Flood Damage Reduction Program (FDRP). It was during this period that flood management strategies shifted towards prevention, modification, and redistribution of flood losses. According to Shrubsole, the current phase in flood management in Canada is influenced by the public's growing awareness of environmental management issues, by the need to respond to the budget cuts of the mid-1990s, and by the withdrawal of Environment Canada from the FDRP.

> Institutional rather than technical factors lie at the heart of improved flood and hazard management in Canada. If improvements to flood and hazard management are to be achieved, institutional arrangements must be reformed. Flood and hazard management cannot be seen in the context of building dams, dykes and channel improvements, the area under regulations or the number of people evacuated; it is about what these measures mean for Canadian society and local communities. This will require a change in the culture and institutional arrangements for flood management at all levels. (Shrubsole 2007, 194)

Most flood-control structures across Canada today are smaller versions of the structures used on the Red and Assiniboine Rivers in Manitoba or on the Fraser River in British Columbia. Although few Canadians will remember the Red River flood of 1950, its legacy remains in the form of the Red River Floodway. The 47 km floodway serves to divert the Red River around the eastern edge of Winnipeg, where it continues on its course to Lake Winnipeg (Figure 5.7). Manitobans had been battling the Red River since 1826, and the construction of the floodway (built at a cost of just over $60 million in the 1960s) was the culmination of decades of efforts to mitigate flooding risk. The floodway has saved the city from flooding 18 times. It is estimated that without the floodway during the 1997 flood, 80 per cent of Winnipeg would have been under water and 500,000 city dwellers evacuated. The floodway is estimated to have saved the city between $6 and $10 billion in damages (Shrubsole 2000).

The Red River Floodway Expansion Project will increase flood protection for the residents of Winnipeg and surrounding areas. Once completed, the project will protect more than 450,000 Manitobans, over 140,000 homes, and over 8000 businesses; it will also prevent more than $12 billion in damages to the provincial economy in the event of a '1-in-700-year' flood. By increasing the capacity of the floodway channel from 1700 m^3 of water per second to 4000 m^3 per second, floodway expansion will increase the level of protection from floods with a 1-in-90-year probability of recurrence to floods with a 1-in-700-year probability (Manitoba Floodway Authority 2008). The complete expansion project is expected to be complete by the spring of 2010.

The irony remains that decisions to reduce expenditures by many levels of government have come at a time when the economic losses caused by flooding are increasing. During the 1990s, Environment Canada withdrew its support from the Federal Flood Damage Reduction Program and no other level of government has filled the void. As a result, a 'cycle of escalating flood losses due to extreme events should be expected from present arrangements' (Shrubsole 2000, 1).

Fog

Fog does not immediately come to mind when one thinks about severe Canadian weather, and yet both in cost and casualties, fog has consistently had a significant impact on society, particularly the transportation sector—sometimes with devastating consequences (Whiffen et al. 2003). On Canada's highways, more than 50 fatal collisions occur each year in dense fog/smog/mist conditions (Transport Canada 2001). From a global perspective, Canada is home to some of the foggiest places on Earth (Figure 5.8).

Within Canada, meteorologists and the aviation community define *fog* as a suspension of very small water droplets that reduces the horizontal visibility to less than 1 km. The fog is defined as 'heavy' when the horizontal visibility is reduced or obscured to less than 0.4 km. In addition, fog may be characterized as dense when zero visibility is reported. In many cases, the occurrence of fog is localized and influenced by factors such as topography, proximity to water bodies, and soil type. As such, the spatial distribution and occurrence of fog are often not captured by the Canadian network of hourly weather observation stations (airports) that report fog. The peak time of year for fog occurrence in Canada varies across the country. The highest probabilities in southwestern British Columbia are near sunrise from August through October. In Vancouver and on the lower BC mainland, the peak probability is about 8 per cent, and it occurs in October. In Whitehorse, the peak time for fog is in January near sunrise. This fog is ice fog in nature, consistent with minimum temperatures around that

Figure 5.7 Red River Floodway.

Source: Reproduced with the permission of Natural Resources Canada 2008, courtesy of the Geological Survey of Canada (Photographer: G.R. Brooks).

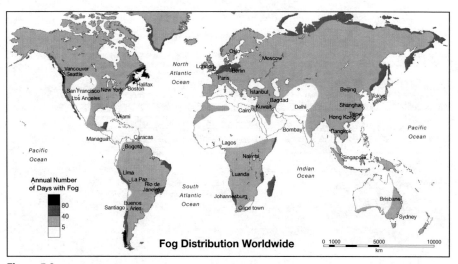

Figure 5.8 Global patterns of fog.
Source: Modified from Burt C., 2004.

time dipping down to −40°C. In Resolute, fog is most frequent in July and August, coincident with the open-water season and the highest average temperatures and highest maximum temperatures, conditions conducive to the formation of advection fog (described below).

The foggiest place in Ontario is North Bay. Across the province, there is a large amount of variability in seasonal fog trends. The peak probability for fog in the Quebec City area is during the two hours after sunrise during the months from April to November. The foggiest time of year in Halifax is June and July, with an average of 15 days with fog in June and 18 days in July (Hanson 2006). Each year, there is an average of 122 days with fog at the International Airport outside of Halifax. No part of Nova Scotia is fog free, although some places inland from the Minas Basin have a fog frequency no greater than Toronto's. Nova Scotia's most persistent spell of fog occurred during Canada's centennial in 1967 at Yarmouth, where 85 of the 92 days of summer had an occurrence of one or more hours with fog (Environment Canada 2007c).

One of the foggiest parts of the country is Newfoundland, where the total number of hours of bright sunshine is usually less than 1600 hours a year, which is well below Summerside's (PEI) average of 1959 hours, Calgary's 2314 hours, and the Canadian average of 1925 hours. The waters off the Avalon Peninsula and over the Grand Banks are among the foggiest in the world. The fogs, sometimes known as 'sea smoke', develop when warm, humid air from the south strikes the cold, sometimes ice-infested waters of the Labrador Current. These fogs may occur in all seasons, but on average they are most frequent in the spring and early summer when the contrast between sea and air temperatures is greatest (anywhere between 5°C and 15°C) (Environment Canada 2006b). Surprisingly, the fogs are often accompanied by strong winds. Normally, winds can be expected to disperse fog, but here the fog is frequently so dense and widespread that the winds have little clearing effect. The resulting conditions can be hazardous for shipping and for drilling rigs, especially when icebergs are present.

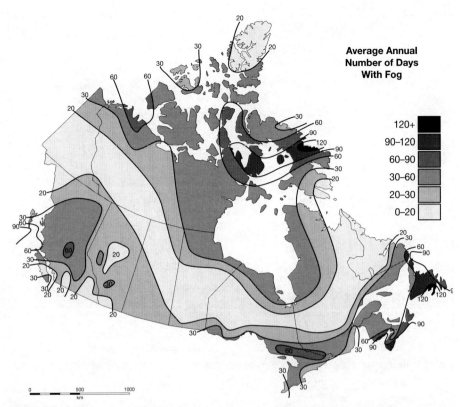

Figure 5.9 Average number of days per year when fog reduces visibility to less than 1 km, based on data from 1971 to1999 (Canadian overview).

Source: Modified from Muraca et al. 2001.

Figure 5.9 shows the average number of days per year in which outbreaks of fog reduce visibility to less than 1 km. Higher occurrences exist along coastal areas (Figure 5.10), in interior British Columbia, and in southern Ontario and the Arctic. Muraca et al. (2001) offer the qualification that the data for this is based on the available, sometimes widely spaced station observations, meaning that locally enhanced fog areas across Canada may not be identified. Results from the 1971–99 fog study undertaken by Muraca et al. (2001) indicate a general decrease in the number of fog days across Canada during the 30-year study period. In fact, this declining fog trend was typical of 95 per cent of stations reporting in Canada, with a reduction in fog events in the order of 20–40 per cent from 1971 to 1999. Other global fog research (e.g., in England and the United States) suggests that this trend towards reduction in fog is not limited to Canada. Further research is required to determine the possible causes of these observed reductions in fog.

Processes and Types of Fog

There are several types of fog, each formed by variations in the cooling/condensation process. Frontal fog and radiation fog and are perhaps the most common fog types in

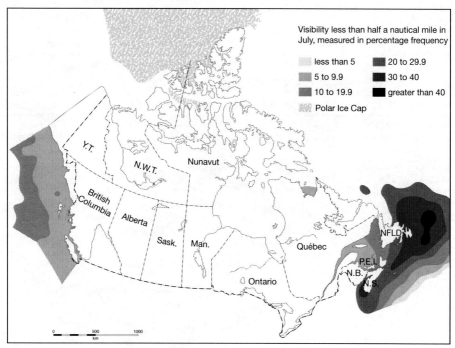

Figure 5.10 Fog-prone areas on Canada's east and west coasts.
Source: Cathy Conrad.

areas away from large water bodies. *Frontal fog* is associated with weather fronts, particularly warm fronts. This type of fog develops when frontal precipitation falling into the colder air ahead of a warm front causes the air to become saturated through evaporation. *Radiation fog* generally forms when the air near the ground surface cools to its saturation temperature owing to radiative cooling at night, when the sun has set. The most studied fog types are those associated with radiative cooling over land. In hilly or mountainous terrain, one might experience valley or upslope fogs. *Valley fogs* involve the mixing of dense, cold, moist air in a valley bottom. Air near the higher elevations cools and descends through its greater density into the surrounding valleys, flowing like water. Pooled in the valley, the cold air may condense the water vapour present into a fog that fills the valley. *Upslope fogs* result from air being forced to rise upslope along a topographic barrier, cooling as it rises. During this cooling, if the air temperature falls below the dew point, the resulting condensation forms a cloud. If the cloud hugs the ground, it becomes fog at that surface (Heidorn 2002).

Advection fog is another relatively well-studied fog type. It is produced when a warm, moist air mass moves over a relatively colder, drier ground surface. The term *advection* means the horizontal movement of air. Unlike radiation fog, advection fog can occur in windy conditions. Also unlike radiation fog, advection fog can occur when the skies aloft are initially cloudy. The set-up for advection fog often includes an advection pattern that brings in warmer and moister air from the south. The primary form of fog in the Maritimes region is due to advection of high dew-point air over water with a cool sea surface temperature (SST). The SSTs around Charlottetown in the Northumberland Strait become warmer faster (20°C to 21°C by late July) than along the Atlantic coast of mainland Nova Scotia. Also, there are frequent cool water upwelling events along Nova Scotia's Atlantic coast.

Many coastal regions of Canada also suffer from *coastal fog*, which forms when moist air is cooled to the saturation point by travelling over a cooler sea. The wind may then take the fog into coastal regions. This type of fog tends to occur in spring and summer, and particularly affects parts of Newfoundland and Nova Scotia (see photo of the Golden Gate Bridge in colour insert section). So-called *sea fog* is a well-known example of advection fog in coastal regions of Canada. Because sea fog typically occurs as a result of warm marine air advection over a region affected by a cold ocean current, it is common at sea in locations where boundaries with cold ocean currents can be found, such as the Grand Banks of Newfoundland (Gultepe et al. 2007).

Impacts of Fog in Canada

Fog has been chosen as one of Environment Canada's annual 'Top Ten Weather Stories' on two occasions. On 3 September 1999, a dense early-morning fog enveloped sections of Highway 401 near Windsor, contributing to one of the worst road disasters in Canadian history. Eight people died, and a total of 82 vehicles were destroyed in the fiery pileup, many having fused together in the intense heat. Moments before the crash, visibility had been reduced to about a metre by the sudden occurrence of dense fog just after sunrise. Because of the small scale of the fog event and lack of *in situ* measurements in the area of the accident, it was difficult to analyse the fog and no definitive answers on its formation could be provided (Pagowski et al. 2004). On 17 March 2003, a series of accidents on Highway 400 near Barrie, Ontario, was blamed on another occurrence of thick morning fog, the result of warm temperatures and rapidly melting snow. Two dozen people were injured in the pileups, which spanned a 20 km stretch of highway and involved over 200 vehicles (Environment Canada 2005a).

Radiation fog in particular creates dangerous driving conditions. Smaller droplet size and larger concentrations per volume of air disrupt the path of light (and therefore affect visibility) more significantly than fog formed from other processes. Radiation fog also forms suddenly, frequently near sunrise, in otherwise ideal conditions (clear skies, light winds), and therefore catches motorists by surprise. The transition from clear conditions to extremely low visibility is much more abrupt with radiation fog than with other types of fog or even other types of weather (Whiffen et al. 2003).

Fog is not only hazardous on land, but also over water (Figure 5.10). The worst maritime disaster in Canadian history was the fog-related collision in May 1914 between the Canadian Pacific's *Empress of Ireland* and the Norwegian coal ship *Storstadt* in the St Lawrence River near Rimouski, Quebec. More than 1000 people died. The captain of the *Empress of Ireland*, H.G. Kendall, told an inquest that he had brought his ship to a halt and was waiting for clearer weather when, to his horror, another ship emerged from the fog, bearing down upon him less than a ship's length away (PCS 2007b). The culprit was fog, but a fog peculiar to the St Lawrence at this time of year, when the warm air of late spring encounters a river chilled by icy meltwater.

The ships sighted each other near 2:00 a.m. on May 29, until then a calm, clear night. On the bridge of the Empress of Ireland, Captain Henry Kendall guessed that the approaching ship was roughly eight miles away, giving him ample time to cross her bow before he set his course for more open water. When he judged he was safely beyond the collier's path, he did so. If he held his new course, the two ships should pass starboard side to starboard side, comfortably apart. Moments after he had executed this maneuver, a creeping bank of fog swallowed the Norwegian ship, then the Empress. (PBS 2007)

Fog can also kill people directly, although fortunately this hasn't happened in Canada. In virtually every case, these killer fogs have been the carriers of industrial atmospheric pollution. For example, 60 Belgians died in 1930 when fog conditions concentrated pollutants in the Meuse Valley for six days. In late October 1948, nearly half the residents of Donora and Webster, Pennsylvania, fell sick, 20 of them dying, as a result of a fog that held industrial pollution in the towns. Thousands of Londoners died in December 1952 when a toxic fog descended on the English capital, a fog so thick that some reported they could not see their feet as they walked the London streets. Carrying a deadly mixture of dense coal smoke, the fog literally poisoned the populace. It lasted from the 5th to the 9th of December, and in the end, as many as 12,000 died from its effects. It is still considered one of the deadliest environmental incidents on record.

Occasionally, fog is deadly without pollutants, although not to humans. A fog bank killed more than 5000 songbirds over the Bay of Fundy between New Brunswick and Nova Scotia in June 2004. The birds apparently died of hypothermia when they flew into the cold fog. When birds get wet, they have difficulty flying and become cold. Cold, wet birds have to stop flying, but unfortunately, in that June 2004 fog, there wasn't land under most of them. Lobster fisherman said the birds were just dropping from the sky, some landing dead on their boats (CBC 2004a).

Because fog can be a significant hazard, there have been attempts to modify foggy weather with fog dispersal techniques. During World War II, the British had considerable success in clearing radiation fog away from runways. Fuel burners were deployed along airfields, raising the air temperature and thereby lowering the relative humidity below the saturation point so that fog droplets vaporized. This technique resulted in the safe landing of an estimated 2500 planes. Although this approach to fog dispersal was successful, the fuel costs associated with it made its broader use prohibitive. Airports around the world have employed other techniques for fog dispersal, including cloud seeding and the use of helicopters to induce vertical air circulation. The Orly and De Gaulle airports outside Paris, France, employ one of the more sophisticated techniques in use today. Their so-called Turboclair system includes underground jet engines that direct streams of warm exhaust over runways, thereby lowering the relative humidity and causing fog droplets to vaporize (Moran and Morgan 1997). There are presently no fog-warning systems in place in Canada, although systems used in the United States could be modified for Canadian communities that are susceptible to particularly foggy weather.

Drought

Droughts are a form of natural hazard that can have devastating effects on the environment, society, and the economy. However, as one of the most complex of weather hazards, they are very difficult to define. The comprehensive definition of a *drought* is a 'prolonged period of abnormally dry weather that depletes water resources for human and environmental needs' (Atmospheric Environment Service Drought Study Group 1986). Other definitions exist for specific categories of drought.

Droughts, at best, are inconvenient, and at worst, they can be catastrophic to regional economies. The three predominant categories of drought are meteorological, agricultural, and hydrological. A *meteorological drought* occurs when seasonal or annual precipitation is below its long-term average; this type is rarely associated with the effects of a reduced water supply on humans or the land. A *hydrological drought* affects bodies of water such as lakes, rivers, or basins, where low levels of precipitation lead to lower water levels over time. This type of drought usually occurs after a prolonged meteorological drought. When the total amount of water accumulating in a water body is less than the amount that is evaporating, a hydrological drought is considered to be in effect. Aquatic species may be adversely affected by such droughts, and water supplies for urban, industrial, residential, and hydroelectric use may be compromised. An *agricultural drought* exists when there is not enough water available to sustain crop growth. The time of year that drought conditions emerge has a decided impact on how damaging the conditions will be for a particular crop. Drought conditions that emerge in the early summer months may have a much greater impact on forage crops and livestock than droughts in late summer, which have a strong impact on cereals and oilseeds.

More recently, there has been reference to a fourth category of drought—*socio-economic*. In September 2006, after experiencing a long, hot summer with no rain (first a meteorological drought and then a hydrological drought), the Vancouver Island resort community of Tofino was suffering through a severe water shortage. Tofino is generally a very wet place, but most of its rain falls over the winter. The town does not have the ability to store all the water that will be needed for the busy summer tourist season, when the community of 1800 swells to 18,000 people. The water shortage in 2006 prompted the community to order all local businesses to close. Tourists and wedding parties were left scrambling to make alternate plans for the last holiday weekend of the summer (CBC 2006l). Ultimately, the town had to truck in 200,000 gallons of municipal water (*Vancouver Sun* 2006).

Droughts are categorized according to the area affected, their timing, and their duration. A drought may affect crop growth or soil erosion (agricultural drought), but not last long enough to impact water supplies (hydrological drought), for example. The extent of the drought's impacts depends on a region's sensitivity, vulnerability, and ability to adapt (Wheaton et al. 2005). Droughts are unique among natural disasters in that their recurrence in the same location every few years is practically certain (Maybank et al. 1995), and they are far from being a recent natural hazard. As far back as the 1790s, there was a drought so severe that it turned the dormant sand dune fields in Alberta and Saskatchewan into active flowing sand.

Lack of rainfall during the years 1930–34 over a large part of the Prairie provinces created an acute shortage in both surface and groundwater supplies. These conditions were exacerbated by land use practices. At that time, farmers broke up the soil more than they do today, making it more vulnerable to erosion and windstorms. The drought brought other problems, including grasshoppers and a weed called Russian thistle. The grasshoppers were so thick that they clogged the radiators of cars and made the roads slippery. Chickens and turkeys ate the insects, which gave a foul taste to their meat and eggs. By 1937, these conditions had reached their peak. There was no hay to feed starving livestock, causing the price of cattle to drop to 3.5 to 4 cents per animal (CBC 2004b). As soil turned to dust and winds swept across the prairie, the dust invaded homes, covering furniture and possessions. The 'Dirty Thirties' became an appropriately descriptive term for this period.

Droughts present a serious national threat to Canada (Bonsal et al. 2004). They occur in all parts of the country, but because they are infrequent in most inhabited areas, there is no national long-range planning program for them. This is not the case for the Canadian Prairies, however, where droughts can have—and have had—a wide range of economic impacts. In part because of Canada's large area, mild or more severe drought conditions are experienced in some part of the country for some part of the year. The impacts of these drought episodes are different for each part of the country.

> For example, forest fires often increase due to droughts in the coastal provinces and in the Canadian North where forest vegetation dominates. Other areas, such as the Okanagan Valley in British Columbia and the St. Lawrence Valley can be affected by agricultural droughts. However . . . the most severe and widespread droughts in Canada occur on the Canadian Prairies. The Prairies, located on the northern extremity of the North American Great Plains, are a semi-arid to sub-humid area that experiences highly variable weather. This is primarily due to its location in the lee of the western cordillera, distant from the moderating influence of the ocean. (Maybank et al. 1995, 198)

Droughts in other parts of Canada sometimes occur in different years than the Prairie droughts. Droughts in the Maritimes have been historically more rare, although there have recently been several dry years in Nova Scotia. In 2006, the Nova Scotia Department of Agriculture and Fisheries hosted the Prairie Farm Rehabilitation Administration (a group that specializes in water resource management) in an effort to find solutions for Nova Scotia's agricultural community. Because droughts occur with less frequency in Atlantic Canada than in many other parts of the country, the area has a lower adaptive capacity and is thus particularly susceptible to impacts when they do occur. Droughts are of less concern in northern Canada, mainly because of their low frequency in the region and lower population densities.

Circulation patterns in the upper atmosphere are a major factor in the onset and perpetuation of droughts. Studies across the Canadian Prairies have been conducted to determine the cause of drought conditions there. Dey (1982) found that droughts over the Canadian Prairies are often associated with quasi-stationary mid-tropospheric ridges that block precipitation. A strong relationship between seasonal Canadian temperatures and precipitation, key factors in the formation of droughts, and the El Niño–Southern Oscillation (ENSO) has also been established (e.g., Shabbar et al. 1997). Bonsal and Lawford (1999) found that the Canadian Prairies tend to experience more frequent summer dry spells and decreased precipitation during the second summer following an El Niño event. Several large North American droughts over the past century have had a tendency to be clustered in successive years and are likely related to ocean-atmosphere variability in similar time scales. The Dust Bowl drought of the 1930s was a natural disaster that severely affected much of the western United States and western Canada, occurring in three waves, in 1934, 1936, and 1939–40 (Shabbar and Skinner 2004).

Most research focuses on the large-scale conditions associated with droughts, but some studies have looked at the small-scale characteristics of the weather. Surprisingly, Roberts et al. (2006) found that there is little change in the amount of total cloud cover during drought conditions, but that cloud bases are higher and the maximum temperatures occur one to two hours later in the afternoon.

Severe Canadian Droughts

Droughts typically occur once every three years somewhere in Canada (Roberts et al. 2006). Almost every decade since records began to be kept in the seventeenth century has included at least one significant drought year. Three decades, 1910–20, 1930–39, and 1980–89, stand out as dry because they were drought-stricken for at least half their length. Nkemdirim and Weber (1999) conclude that although droughts were more frequent and persistent in the 1930s than in the 1980s, the droughts of the 1980s were more severe and spanned larger areas. Whereas the pattern of severity was patch-like during the 1930s, large continuous drought zones marked the events of the 1980s. In the 1980s, a series of highs blocked out storms, thus creating a regional drought pattern.

Droughts in Canada usually affect only one or two regions, tend to be relatively short-lived (one or two seasons), and have an impact on only a few sectors of the economy. There are certainly exceptions to this general description, however. Some regions of southern Alberta and Saskatchewan experienced drought conditions for as many as eight consecutive years. During the 1950s, the Canadian Prairies were subjected to a five-year drought, and in three of those years, drought conditions stretched from coast to coast. The 1950s drought was characterized by both below-normal rainfall and higher-than-normal air temperatures (Shabbar and Skinner 2004).

Between 1900 and 1993, there were 38 years during which the Canadian Prairies were afflicted by droughts severe enough to have a significant effect on crop production (Roberts et al. 2006). The events that have been considered the most severe are the multi-year droughts of the 1930s and the 1980s and the single-year drought of 1961. A comparison among these three major drought events shows that severe droughts can differ significantly in terms of timing, areal extent, duration, and drought type itself, as well as in terms of environmental and socio-economic effects (Maybank et al. 1995). Drought-related losses were in excess of $1.8 billion in 1988 (Shabbar and Skinner 2004). The droughts of 2001 and 2002 covered massive areas of Canada, were long-lasting, and brought conditions that had not been seen for at least a hundred years in some regions. The drought of 2001–2 was rare in that it struck areas not usually affected by drought, such as eastern Canada and the northern Prairies. The drought of 2001–2 hit Saskatchewan and Alberta the hardest, however. Agriculture losses were estimated at more than $2 billion, and employment losses exceeded 41,000 jobs (Wheaton et al. 2005). Shabbar and Skinner (2004) report a total of $6.14 billion in economic losses due to the 2001–2 drought. As severe as these droughts were, the 1961 drought remains the single most severe growing-season drought to occur on the Canadian Prairies.

The 1930s have come to be known as the 'Dirty Thirties' thanks to the giant dust storms that swept across the North American prairies. The cause of the dust storms was a combination of poor farming practices and subsequent drought conditions. The heavily tilled prairie grasslands gave way to loosened soil that was prone to wind erosion. The heavy equipment that farmers had begun using in the 1920s contributed to the baring of more and more land. When a drought broke out across North America, the flatness of the prairie and the sparse tree cover provided little resistance to the wind. As vast quantities of the rich prairie topsoil were blown away, huge clouds of black dust formed over the region and travelled as far away as the Atlantic Ocean. Drought conditions have never since been as bad as those in the 1930s, in part because of the lessons learned from that period.

Monitoring and Drought Indices

Agriculture and Agri-Food Canada oversees 'Drought Watch', the principal goals of which are to provide timely information on the impacts of climatic variability on water supply and agriculture, and to promote practices that reduce drought vulnerability and improve drought management. Using the Drought Watch website, the public can utilize maps of current and long-term conditions of precipitation (see map of Moisture Anomaly Index in colour insert section).

The North American Drought Monitor (NADM) is a cooperative effort undertaken by drought experts in Canada, Mexico, and the United States to monitor drought across the continent on an ongoing basis (NOAA 2007c). Drought monitoring is an integral part of drought planning, preparedness, and mitigation efforts, with costs and effects of drought extending beyond international borders. The NADM initiative is in the process of developing continent-scale databases, creating product displays, and working out the logistics of continent-scale map displays (Lawrimore et al. 2002).

The uncertainty of drought prediction contributes to substantial crop insurance payouts every year. A number of different indices have been developed to quantify drought, each with its own strengths and weaknesses. The Palmer Drought Severity Index (PDSI) and the Standard Precipitation Index (SPI) are two of the most commonly used indices. The PDSI, the most frequently cited index (Shabbar and Skinner 2004), reflects long-term moisture, runoff, recharge, deep percolation, and evaporation. It is useful for drought analysis over time spans of months or seasons. Maps of monthly PDSI patterns for Canada can be found on the Drought Watch website. The map 'Seven Day Accumulated Precipitation' in the colour insert section provides an example of the spatial distribution of the PDSI values for the agricultural land across Canada for the month of September 2006. The index uses dimensionless numbers, typically ranging from −4 to 4, with negative quantities indicating a shortage of water, while positive numbers suggest surplus water. This example of the PDSI reflects a combination of the moisture conditions for the current month and the moisture conditions for the previous months (Agriculture and Agri-Food Canada 2007). It shows a dry region in the vicinity of Edmonton, an area north of Winnipeg and along the central-eastern region of British Columbia. Table 5.4 provides a qualitative description associated with each of the Palmer Index values.

Like the PDSI, the Standard Precipitation Index assists in the forecasting and monitoring of droughts. The SPI was created in 1993 by researchers at the Colorado Climate Center to measure 'the difference of precipitation from the mean for a specified time divided by the standard deviation, where the mean and standard deviation are determined from the climatological record' (AMS 2008). The SPI was designed to have an advantage over the Palmer Index in that it can quantify precipitation deficits for different time scales (Guttman 1998).

Some researchers have concluded that, beyond the initiatives that have been undertaken in terms of drought monitoring, we must develop the capacity to predict droughts as part of seasonal forecasting efforts. More detailed studies must be undertaken to help us better understand all aspects of Canadian droughts (Wheaton et al. 2005).

Climate Change and Precipitation in Canada

The 2007 report of the Intergovernmental Panel on Climate Change (IPCC 2007b) concludes that the global frequency of heavy precipitation events has increased over

Table 5.4 The qualitative explanations for the Palmer Drought Severity Index values.
Source: Modified from Agriculture and Agri-Food Canada 2007.

PDSI Values	Category
5.00 or greater	Extremely wet
4.00 to 4.99	Severely wet
3.00 to 3.99	Moderately wet
2.00 to 2.99	Slightly wet
1.00 to 1.99	Incipiently wet
−0.99 to 0.99	Near normal
−1.99 to −1.00	Incipiently dry
−2.99 to −2.00	Mildly dry
−3.99 to −3.00	Moderately dry
−4.99 to −4.00	Severely dry
−5.00 or less	Extremely dry

most land areas and that there is a 90–99 per cent chance that heavy precipitation events will increase in frequency over most areas into the twenty-first century. By the end of the twenty-first century, parts of Canada might expect a 10–20 per cent increase in precipitation (above the 1961–90 average baseline for precipitation). As with many studies related to the impacts of climate change, the IPCC report reflects a certain degree of debate, although even amidst the debate, there have been significant findings over the past few decades.

Karl and Easterling (1999) found that an increasing share of US precipitation was falling in more intense bursts. This finding is in accord with the idea that a warmer atmosphere would carry more water vapour, which in turn would lead to heavier rainfall events. In a mid-latitude country like Canada, heavy rainfalls (and snowfalls) are usually associated with storms and high winds. If extreme precipitation increases, then presumably severe storms would as well.

Zhang et al. (2001) examined the spatial and temporal trends of heavy precipitation across Canada from 1900 to 1998. They found that decadal variability is the dominant feature in both the frequency and the intensity of extreme precipitation events across the country (i.e., wetter decades followed by drier decades). For the country as a whole, there do not appear to have been any identifiable trends in extreme precipitation in the last century. The observed 12 per cent increase in total precipitation (primarily in southern Canada) in the twentieth century was mainly due to an increase in the number of small to moderate events (Zhang et al. 2000). However, heavy rainfall events over eastern Canada in the spring have shown an increasing trend superimposed on the strong decadal variability. Using historical weather data from the three Canadian Prairie provinces, Akinremi et al. (1999) found that precipitation increased about 10 per cent between 1921 and 1995. On the other hand, Ripley (1986), after examining annual and seasonal precipitation trends at three stations in Saskatchewan, notes a decrease in summer precipitation and an increase in spring and winter precipitation. Akinremi and McGinn (2001) found a significant increase in both the rainfall amount and the number of rainfall events on the Canadian Prairies within the 40-year period from 1956 to 1995. Annual rainfall had increased about 16 per cent, while the number of rainfall events increased about 29 per cent. These increases were limited to the first eight months of the year and were not uniform across the Prairies. Stone et al. (2000) found that total annual precipitation increased in southern Canada during the twentieth century (record lengths ranging from 34 to 102 years within the 1895–1996 period) partly due to an increase in intermediate and heavy precipitation events. Mekis and Hogg (1999) report that there was a significant rise in total annual precipitation in almost all regions of Canada over the twentieth century. It remains to be seen what the trends into the middle of the twenty-first century have to hold for regions across Canada, as there does not appear to be a clear consensus on the trends in the recent past.

Zhang et al. (2000) have determined that, on average, there are 20.3 more days with rain in Canada today than in the 1950s, but only an increase of 1.8 days with heavy precipitation. The results of their study suggest that the increase in annual total precipitation during the second half of the twentieth century was due mostly to more days with precipitation, with no consistent changes having been observed in extreme precipitation. The lack of consistent patterns in extreme precipitation highlights the need for regional studies of the local characteristics of precipitation extremes. Stone et al. (2000) found that the increase in precipitation in the summer in the southeast of Canada during the last half of the twentieth century was related to increasingly frequent heavy rainfall events, with the same trends occurring in the Arctic in all seasons but the summer. Ashmore and Church (2001) suggest that the general increase in the magnitude of flood events along many Canadian rivers in this period relative to the first half of the century may reflect a shift in climate. Although rainstorms and urban floods are likely to become more intense if precipitation intensities increase, 'there is no indisputable evidence from either trends or models to indicate that extreme flood and rainfall events in Canada are either increasing or decreasing in magnitude or frequency' (Lawford et al. 1995, 324).

Natural Resources Canada (2006) reports, however, that flood disasters have increased in frequency through the twentieth century, with about 70 per cent occurring after 1959. The increase must be qualified, however, as it is indicative of more than simply an increase in environmental processes. One could argue that the extent of a disaster is at least in part attributable to the increase in the human vulnerability to such events (more infrastructure, a greater percentage of built environment, and larger numbers of people residing in flood-prone areas as Canada's population has grown). Furthermore, the reporting of flood disasters has improved over the past several decades. It is likely the case that smaller pre-1960 flood disasters are under-represented in the database because they were usually only reported in small local papers and were thus overlooked as 'disasters' (Natural Resources Canada 2006). Kunkel (2003) raises the important point that flooding can be caused by increased frequency/intensity of precipitation in combination with human alterations in land use and stream-channel modification. This situation has clouded the issue of whether floods are increasing and whether such trends are solely or largely due to weather changes.

Climate Change and Fog

Several studies have revealed a widespread decrease in low visibility caused by fog, with long-term decreases in fog over several decades reported in Canada, Argentina, and Brazil. Fog is generally not a major topic in climate change studies, but it could very well play a part. Over the past 30 years, 20–30 per cent fewer fog events have been reported in Canada. For example, from 1951 to 1980, some Atlantic coast stations reported an average of over 150 days of fog per year, while from 1971 through 1999, that number dropped to a maximum of 132 days per year (Muraca et al. 2001). Because fog is influenced by so many factors spanning multiple spatial and temporal scales, significant difficulties in its accurate and detailed short-term prediction persist (Whiffen et al. 2003; Gultepe et al. 2007).

Climate Change and Implications for Drought

Evidence indicates that the risk of agricultural and socio-economic drought is increasing as demands for food and water increase and the manifestations of climate change become ever more apparent (Wheaton et al. 2005). According to Maybank et al. (1995), the frequency and intensity of droughts will increase but this could be offset by less variability in precipitation. More recent studies indicate that the latter may be the case. The maximum length of consecutive dry days in Canada has decreased by about 10 days since the beginning of the twentieth century, with a shift towards shorter dry periods in the second half of the century (Zhang et al. 2000). In spite of the general increase in precipitation throughout the century, the area most affected by abnormally warm and dry conditions has remained constant in the summer months and has even increased in the important spring growing season (Zhang et al. 2000). With the largest increases in precipitation in Alberta and Saskatchewan in the latter half of the twentieth century, it would not appear that the Prairies are getting drier (Akinremi and McGinn 2001).

More recently, there has been some effort to define large-scale trends and variability in Canadian temperatures, precipitation, and drought-related indices (e.g., Bonsal et al. 2004). The results show substantial decadal variability with no consistent trends in terms of the frequency, duration, or severity of droughts during the twentieth century. Unfortunately, most studies employ different temporal and spatial scales and are limited in terms of data quantity and quality.

> All global climate models are projecting future increases of summer continental interior drying and associated risk of droughts. The increased drought risk is ascribed to a combination of increased temperature and potential evaporation not being balanced by precipitation. However, considerable uncertainty exists with respect to future precipitation, particularly on a regional and intra-seasonal basis. Furthermore, relatively little is known regarding changes to large-scale circulation and, since these patterns have a significant impact on temperature and precipitation over Canada, the occurrence of future drought remains a huge knowledge gap. (Bonsal et al. 2004, 23)

The potential impacts of climate change across Canada are identified in a study conducted by Natural Resources Canada (2007). Water shortages are noted to have been documented in southern regions of Ontario and are projected to become more frequent as summer temperatures and evaporation rates increase. Increases in water scarcity represent the most serious climate risk in the Prairie provinces. Communities dependent on agriculture and forestry are highly sensitive to climate variability and extremes. Droughts (with the associated economic impacts in the billions of dollars), wildfires, and severe floods are projected to occur more frequently in the future. Many regions and sectors of British Columbia will experience increasing water shortages and more competition among water uses (e.g., hydroelectricity, irrigation, community needs, and recreation needs), with implications for transborder agreements. British Columbia's agricultural sector faces both positive and negative impacts from climate change, with more frequent and more sustained droughts being the greatest risk.

Chapter Summary

Large amounts of precipitation fall in regions of British Columbia and Yukon, most heavily on the western side of the mountain ranges. The Prairies and the Arctic regions emerge as relatively dry regions in the spatial patterns. Moderate amounts of precipitation fall throughout Ontario, Quebec, and the Atlantic provinces. Flooding in Canada can occur virtually anywhere, any time of year. Some areas have been historically more vulnerable than others. In central and eastern Canada, flood disasters have primarily occurred in the south, where the population is concentrated. Flooding conditions occur as a consequence of one or a combination of the following factors: snowmelt runoff, heavy rainfall, ice jams, catastrophic outburst, coastal storms, urban stormwater runoff, and dam failure. The rapid expansion of Canada's population during the 1950s and 1960s gave rise to significant urban development, considerably increasing the population's exposure to flooding. Most flood-control structures across Canada today are smaller versions of the structures used on the Red and Assiniboine Rivers in Manitoba and on the Fraser River in British Columbia.

The occurrence of fog across Canada is localized and influenced by factors such as topography, proximity to water bodies, and soil type. Coastal areas of Canada experience the largest number of fog conditions, with the Atlantic provinces being among some of the foggiest places on Earth. Fog does not immediately come to mind when one thinks about severe Canadian weather, and yet both in cost and casualties, on land and at sea, fog has consistently affected society, and in particular the transportation sector—sometimes with devastating consequences.

Meteorological, agricultural, hydrological, and socio-economic droughts have afflicted a variety of regions across Canada and present a serious national threat. In the Canadian Prairies, droughts have had a wide range of economic impacts. Considerable debate remains over exactly how the effects of climate change will be felt across Canada, although there is some consensus building that there may be more precipitation in the form of heavy precipitation events. With larger populations residing in more drought-prone areas, however, increased vulnerability to agricultural and socio-economic droughts may offset such trends.

Chapter Six

Tropical Cyclones

Objectives

- To explain the processes involved in the tropical cyclones and post-tropical and extratropical storm systems that affect parts of Canada
- To describe how tropical cyclones are categorized and named
- To describe and account for the effects of tropical cyclones
- To provide information about the Canadian Hurricane Centre and its tropical storm forecasting
- To describe notable Canadian tropical storms
- To explain the present state of knowledge about the links between climate change and tropical storm systems

Introduction

Although hurricanes (also referred to as tropical cyclones) often bypass eastern and western Canada, moving out to sea, or strike with less intensity, it is important to realize that we are hit by hurricanes and tropical storms every year (Environment Canada 2007b). Because historically it has been less likely that a severe hurricane would strike Canada than the United States or the Caribbean, the general public would likely be surprised to learn that the area near the Atlantic provinces actually has the highest climatological frequency of extratropical events[1] in the North Atlantic (Abraham et al. 2004). Figure 6.1 illustrates the frequency of these events from 1851 to 2003.

---

[1] *Extratropical* is a term used in advisories and tropical summaries to indicate that a cyclone has lost its 'tropical' characteristics. The term implies both poleward displacement of the cyclone and the conversion of the cyclone's primary energy source from the release of latent heat of condensation to baroclinic (the temperature contrast between warm and cold air masses) processes. It is important to note that cyclones can become extratropical and still retain winds of hurricane or tropical storm force.

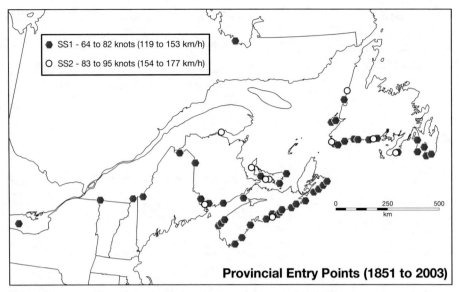

Figure 6.1 Storms of tropical origin with hurricane force winds (1851–2003).
Source: Modified from Environment Canada 2005c.

Because of cooler sea surface temperatures (SSTs) immediately offshore, storms have usually weakened by the time they hit Canada. Warmer SSTs enable hurricanes to continue to build strength, while cooler water temperatures act to mitigate the intensification process. Occasionally, when the warm Gulf Stream extends close to Atlantic Canada, hurricanes strike the eastern provinces with full force (see images of Hurricane Bob and The Gulf Stream in colour insert section). This generally occurs during the latter part of the hurricane season, with the most common statistical date for a landfalling hurricane in Canada being 15 September (Environment Canada 2004i). Because cooler waters extend for a great distance off the Pacific coast of Canada, no tropical cyclone of any intensity has ever struck western Canada directly. Such storms have affected the west indirectly, however. The most notable example is Typhoon Freda, which was absorbed by the Columbus Day Storm of 1962. In recent decades, a reanalysis of the classification of former hurricanes has led to the reclassification of some systems. All references to hurricane strengths and statistics in this chapter relate to information as understood in 2007. There may be further reclassification in the future.

While there is no recorded history of a category 3, 4, or 5 hurricane ever having made landfall in Canada (Bowyer 2007), an average of four tropical cyclones affect eastern Canada every year. Statistically, Newfoundland has the highest probability of being struck by a tropical cyclone (Table 6.1). Tropical cyclones do not strike Quebec and Ontario directly, but rather are always the remnants of tropical cyclones. Although these storms are generally weak and do not cause much damage, there have been notable exceptions, such as Hurricane Hazel (1954).

> **Table 6.1** Frequency of impacts from tropical cyclones by province (1901–2000)
> Source: Modified from Environment Canada 2005c. © Her Majesty the Queen in Right of Canada, 2005. Reproduced with the permission of the Minister of Public Works and Government Services of Canada.
>
> Frequency of tropical cyclones by province:
>
> - Newfoundland: One every 1.68 years
> - Nova Scotia: One every 1.98 years
> - Quebec: One every 2.9 years
> - New Brunswick: One every 3.0 years.
> - Prince Edward Island: One every 5.29 years.
> - Ontario: One every 5.29 years

Processes

Different terms are applied to tropical systems, only some of which will eventually develop into a full-blown hurricane. *Tropical cyclone* is the generic term for the class of tropical weather systems that includes tropical depressions, tropical storms, and hurricanes. Any closed atmospheric circulation system in the Northern Hemisphere in which winds rotate in a counter-clockwise direction is called a cyclone (Barks and Richards 1986).

A tropical cyclone is defined as 'a warm core, non-frontal low pressure system of synoptic scale developing over tropical or subtropical waters' (USDOC 1985) typically around 27°C or warmer; it has organized deep convection and a closed surface-wind circulation about a well-defined centre. Those that affect Atlantic Canada and coastal waters originate in tropical and subtropical latitudes of the North Atlantic Ocean, the Caribbean Sea, and the Gulf of Mexico. Once formed, a tropical cyclone is maintained by the extraction of heat energy from the ocean.

A *post-tropical storm* is a tropical storm or hurricane that moves beyond the tropics into the mid-latitudes and begins to lose its tropical characteristics. The size of the circulation usually expands, the speed of the maximum wind decreases, and the distribution of winds, rainfall, and temperatures becomes more asymmetric (Environment Canada 2004b).

The term *extratropical storm* is used in advisories and tropical summaries to indicate that a cyclone has lost its tropical characteristics. It is used in US National Hurricane Center bulletins and usually signals that the transition from tropical to extratropical is complete. To avoid confusing the public, however, the Canadian Hurricane Centre (CHC) uses the term *post-tropical* in a storm's name even after the storm has become extratropical (Bowyer 2007). The World Meteorological Organization (WMO) recognizes both terms; it defines an extratropical cyclone as 'any cyclonic-scale storm that is not a tropical cyclone, usually referring only to the migrating frontal cyclone of middle and high latitudes.' These extratropical cyclones can still retain winds of hurricane or tropical storm force, however. Most storms of tropical origin are in the process of becoming extratropical as they move over land or into Canada's offshore areas. A cyclone is no less threatening just because it has ceased to be purely tropical. Wind and precipitation distributions are no longer concentrated about the cyclone's circulation centre, but have expanded in area and become asymmetric. This allows for a significant increase in the radii of gale- and hurricane-force winds in an extratropical or transitioning cyclone. Once a tropical cyclone has completed extratropical transition (ET), its radius of influence can increase from roughly 111 km to 300 km (Szymczak and Krishnamurti 2006). A transitioning cyclone can have a devastating impact on Canada. Not only will a larger area be hit by hurricane- or gale-force winds, but the heaviest area of precipitation can be directly over land, even if the extratropical cyclone itself does not make landfall. Approximately 46 per cent of tropical cyclones over the Atlantic Ocean undergo extratropical transition (Abraham et al. 2004).

A *subtropical cyclone* is a non-frontal low-pressure system that has the characteristics of both tropical and extratropical cyclones. In comparison to a tropical cyclone, this system has a relatively broad zone of maximum winds located farther from the centre and, typically, a less symmetric wind field (Environment Canada 2005c). Often, these storms have a radius of maximum winds extending from 100 to 200 km from the centre (typically larger than what is observed for purely tropical systems). Additionally, the maximum sustained winds for subtropical cyclones have not been observed to be stronger than about 119 km/hr (NOAA 2008).

Winds normally increase towards the centre of tropical cyclones, with sustained speeds often exceeding 100 knots (185 km/hr) just outside the centre. Occasionally, sustained winds exceeding 150 knots (278 km/hr) may occur in well-developed systems. The term *sustained wind* refers to the wind speed averaged over a one-minute period. A unique feature of tropical cyclones is the central 'eye'. The *eye* is generally an area of light winds, minimum cloud cover, and minimum sea-level pressure (Neumann et al. 1978; NOAA 1999). The average diameter of an eye is about 30–60 km, but they can range from 8 km to over 200 km (NOAA 2008). This centre of the system is relatively calm, and people in the midst of a hurricane are often amazed when, as the eye moves over, the ferocious wind and rain suddenly stop and the sky clears, and then, just as quickly, the wind and rain return, but this time from the opposite direction. Tropical cyclones display a cloud structure quite different from the warm and cold frontal cloud patterns characteristic of mid-latitude cyclones. Unlike mid-latitude cyclones, which tend to have a comma structure in their mature stages, tropical cyclones are characterized by a circular wall of thick clouds surrounding an eye, with bands of clouds radiating away from the centre of the storm. Clear areas are often visible between these spiral bands (Barks and Richards 1986).

Tropical cyclones are able to reach as far north as the Canadian provinces owing, in part, to the Gulf Stream. The Gulf Stream is a warm water current that originates in the Gulf of Mexico, flows north past the eastern United States towards Newfoundland, and then flows eastwards towards Europe as the North Atlantic Drift. The Gulf Stream extends farther north in late summer and early fall as sea surface temperatures in the Atlantic Ocean increase. This northward transport of warm water allows tropical cyclones to track farther north without losing their warm water energy source. These tropical cyclones usually make landfall in the United States and end up as mainly rain events in the Canadian provinces, although they can still be quite destructive (Szymczak and Krishnamurti 2006). Tropical cyclones show a tendency to follow preferred tracks, depending on the time of year and location. Early-season cyclones (June and July) tend to form in the general area of the Gulf of Mexico and the western

Caribbean. These cyclones are less likely to reach Atlantic Canada than those that originate later in the year (Barks and Richards 1986).

The tracks of cyclones that typically affect Atlantic Canada begin on the south side of the Bermuda-Azores high-pressure area in the broad region from Florida southeast to the Cape Verde Islands. The movement of the cyclones is first in a generally westerly to northwesterly direction under the influence of the tropical easterly winds. Tropical cyclones moving northwest between Bermuda and the southeastern United States gradually come under the influence of the mid-latitude westerly winds. They then curve northeastward and accelerate (Galbraith 1980).

Classification

Tropical cyclones are classified on the basis of the maximum sustained wind speed in the storm. The terms *tropical depression, tropical storm,* and *hurricane* are assigned depending on whether the maximum sustained winds are less than 34, 34 to 63, or greater than 63 knots, respectively (Neumann et al. 1978). Some tropical cyclones have been identified separately as *subtropical cyclones* largely because of their origin at more northerly latitudes. The same classification by wind speed applies. Once a cyclone exceeds 63 knots (117 km/hr), it is classified as a hurricane and is then rated on the Saffir-Simpson Hurricane Scale. This scale gives a 1–5 rating of the hurricane's present intensity and is used to estimate the potential property damage and flooding along the coast with landfall. Wind speed is the determining factor in the scale (NOAA 2007e). It should be kept in mind that severe impacts can occur regardless of the Saffir-Simpson Scale ranking. Most storms that have a severe impact on Canada are not even a '1' on the scale. Hurricane Gabrielle of 2001, for example, which was downgraded to a post-tropical storm by the time it was moving past Newfoundland, caused considerable damage. In St John's, the rainfall caused severe flooding, and the mayor of the city called Gabrielle 'the worst storm in 100 years' (CHC 2002b). The flooding in the city washed out roads and parking lots and filled some basements with several feet (1 m) of water. There were several reports of sewers being unable to accommodate the excessive amounts of water. Hurricane-force wind gusts led to cancelled flights and left thousands without power, telephone, or heat. The passage of Gabrielle caused several million dollars in damage to hundreds of homes and buildings. Table 6.2 provides a list of the category 1 and 2 hurricanes that have struck parts of Canada. No category 3 (or above) storms have ever hit Canada directly.

Category 1 Hurricane: Winds of 64–82 knots or 119–153 km/hr. Storm surge is generally 1.2–1.5 m above normal. No real damage is done to building structures. Damage is primarily to unanchored mobile homes, shrubbery, and trees. Some damage to poorly constructed signs. Also, some coastal road flooding and minor pier damage.

Category 2 Hurricane: Winds of 83–95 knots or 154–177 km/hr. Storm surge is generally 1.8–2.4 m above normal. Some roofing material, door, and window damage of buildings. Considerable damage to shrubbery and trees, with some trees blown down. Considerable damage to mobile homes, poorly constructed signs, and piers. Small craft in unprotected anchorages break moorings.

Table 6.2 Some of the worst tropical cyclones in Canadian history.
Source: Modified from Environment Canada Centre 2007b; Bowyer 2007.

Year	Name	Category (Saffir-Simpson)	Month	Location
1887	1887E	1	8	NL
1891	1891D	1	9	NS, NL
1891	1891I	1	10	PEI, NS, NL
1892	1892B	1	8	NL
1893	1893C	1	8	NS
1893	1893D	1	8	QC, NL
1893	1893E	1	8	NL
1896	1896E	1	10	NS
1908	1908B	1	8	NS
1924	1924B	1	8	PEI, NS, NL
1924	1924C	1	9	NS
1926	1926B	1	8	NS, NL
1927	1927A	2	8	NS, NL
1935	1935A	1	8	NL
1936	1936O	1	9	NS
1937	1937G	1	9	PEI, NL, NS
1940	1940E	1	9	NL, NS
1958	Helene	1	9	NL
1962	Daisy	1	10	NS
1963	Ginny	2	10	PEI, NS, NB
1966	Celia	1	7	QC, NL, NS, NB
1969	Gerda	1	9	QC, NL
1971	Beth	1	8	NS, NL
1995	Luis	1	9	NL
1996	Hortense	1	9	NS
2000	Michael	1	8	NL
2002	Gustav	1	9	NS
2003	Juan	2	9	NS, PEI

Category 3 (Major) Hurricane: Winds of 96–113 knots or 178–209 km/hr. Storm surge is generally 2.7–3.6 m above normal. Some structural damage to small residences and utility buildings. Damage to shrubbery and trees, with foliage blown off and large trees blown down. Mobile homes and poorly constructed signs are destroyed. Flooding near the coast destroys smaller structures; battering from floating debris damages larger structures. Terrain that is continuously lower than 1.5 m above mean sea level may be flooded inland 13 km or more. As noted above, to date Canada has not been

struck directly by any category 3 hurricanes, although some category 2 hurricanes (such as Juan) have had impacts that were more in line with category 3-strength storms (Bowyer 2007).

Category 4 (Major) Hurricane: Winds of 114–135 knots or 210–249 km/hr. Storm surge generally 4–5.5 m above normal. More extensive failures, including some complete roof structure failures on small residences. Shrubs, trees, and all signs are blown down. Complete destruction of mobile homes. Extensive damage to doors and windows. Major damage to lower floors of structures near the shore. Terrain lower than 3 m above sea level may be flooded, requiring massive evacuation of residential areas as far inland as 10 km. To date, Canada has not been hit by a category 4 hurricane.

Category 5 (Major) Hurricane: Winds greater than 135 knots or 249 km/hr. Storm surge generally greater than 5.5 m above normal. Complete roof failure on many residences and industrial buildings. Some complete building failures, with small utility buildings blown over or away. All shrubs, trees, and signs blown down. Complete destruction of mobile homes. Severe and extensive window and door damage. Low-lying escape routes are cut by rising water three to five hours before arrival of the centre of the hurricane. Major damage to lower floors of all structures located less than 4.6 m above sea level and within 0.5 km of the shoreline. Massive evacuation of residential areas on low ground within 8–16 km of the shoreline may be required. To date, no category 5 hurricane has ever struck Canada.

There have been efforts in recent decades to undertake a detailed reanalysis of hurricanes that occurred between 1851 and the present. The Atlantic Basin hurricane database (HURDAT) reanalysis project involves the revision, if necessary, of official tracks and intensities (Landsea et al. 2004).

Naming Hurricanes

Forecasters began naming hurricanes and tropical storms in order to improve their communication with the general public with respect to forecasts, watches, and warnings. By naming hurricanes, they were able to reduce the confusion about what storm was being described. Prior to 1950, hurricanes were named for the year in which they occurred plus a letter from the alphabet (e.g., 1942A, 1942B, etc.) (CHC 2006). Since 1953, Atlantic tropical storms have been given human names from lists created by the National Hurricane Center. The original lists featured only women's names, a fact that has some historical context. An Australian meteorologist had given tropical storms women's names in the late nineteenth century, and during World War II, the military and weather forecasters tended to use female names for storms at sea (NOAA 2007g).

In 1979, men's names were introduced, and today they alternate with women's names. The letters Q, U, X, Y, and Z are not included because there are not many names beginning with those letters. The name lists are maintained and updated by an international committee of the World Meteorological Organization. Six lists of alphabetized names are used in rotation. The 2008 list, for example, will be used again in 2014, with the exception of any 'retired' names that might otherwise be used within that year. If a storm is so deadly or costly that the future use of its name on a different storm would be considered insensitive, then at an annual meeting of a committee of the WMO (called primarily to discuss many other issues) the offending name is removed from the list and another name is selected to replace it. Several names have been changed since the lists were created. For example, on the 2004 list (which will be used again in 2010), Gaston has replaced Georges and Matthew has replaced Mitch. Sixty-seven names have been retired for Atlantic storms to date (2008). From that list, only three have directly affected Canada (Hazel [1954], Edna [1955], and Juan [2003]).

In the event that more than 21 named tropical cyclones occur in the Atlantic Basin in a season, additional storms will take names from the Greek alphabet: Alpha, Beta, Gamma, Delta, and so on. The list went up to Hurricane Epsilon in 2005, a particularly active year that spawned such notorious hurricanes as Katrina.

Impacts of Tropical Cyclones

Apart from their high winds, tropical cyclones are also characterized by such destructive features as torrential rains, storm surges, and embedded tornadoes (particularly at latitudes south of 40°N). Although most of the impacts associated with tropical storms and hurricanes occur in the ocean and coastal areas, these storms can devastate inland areas. The biggest threat to life and property inland is damage from flash flooding caused by excessive rainfall. One of the best-remembered Canadian hurricanes is Hurricane Hazel of 1954, during which over 200 mm of rain fell in less than 24 hours, triggering flash floods that killed 81 people in southern Ontario (CHC 2003).

Wind

> 'It's a night I'll never forget. It was so black and the wind, and the mist: it was like a Stephen King movie. The sound was like a 747 when they've got her revved up, just before takeoff—a high-pitched wailing scream. That's just what it sounded like.'
>
> Angus MacDonald, as quoted in McLeod 2003

The high winds of a tropical cyclone are hazardous for shipping and boating offshore as well on for structures on land. The power of hurricane winds offshore was made apparent in 1989, when Hurricane Gabrielle generated extremely high surf. Even though the eye of Gabrielle was several hundred kilometres away and light winds and clear skies prevailed, beaches in Nova Scotia were closed because of the high surf. The strong winds of hurricanes and tropical storms can cause significant damage once they reach coastal areas. When a hurricane makes landfall or approaches a coast, its winds, aside from producing storm surges, can destroy houses, buildings, and other structures, and the associated flying debris and falling structures can cause injury and death (CHC 2003). In early November 2007, post-tropical Noel hit eastern Canada. Wind gusts reached 180 km/hr at Wreckhouse, Newfoundland, and 135 km/hr at McNabs Island in Halifax Harbour (Environment Canada 2007h). While the advance warnings allowed people time to make extensive preparations that helped minimize the impact of the powerful system, there was still widespread damage. Power outages were reported for over 190,000 residences in eastern Canada, mostly in Nova Scotia. Downed trees, smashed vehicles, and battered homes were reported across the Halifax area. The ocean waves associated with Noel were among the most powerful waves witnessed along the Atlantic coast of Nova Scotia in decades (Bowyer 2008).

It has been noted that in storms that strike New England and the Maritimes, the northerly winds to the right of the storm track are light compared to the often very strong southerly winds to the left of the track (Galbraith 1980). Upon landfall, there is a tendency for friction to reduce the surface wind speeds, the magnitude of the effect being a function of topography. Damaging winds may extend into a land area well in advance of the eye making landfall. Forecasters often use as a rule of thumb the maxim that, following landfall, tropical cyclone winds will decrease to half their value every 10–11 hours (e.g., a 135-knot category 5 hurricane would be a marginal category 1 hurricane within 12 hours of landfall and would be down to a marginal tropical storm within 12 hours after that) (Bowyer 2008). If the storm moves over water again, the circulation at low levels can recover significantly. This effect is significant in the Maritimes, for example, where a storm may rapidly cross Nova Scotia and New Brunswick and then intensify as it moves over the Gulf of St Lawrence (Galbraith 1980).

Winds are light in the eye of a tropical cyclone. If a storm is moving rapidly, as is often the case in the Atlantic region, the eye will tend to elongate in the direction of motion. Eyes have been reported as being as long as 140 km (Dunn and Miller 1960). They generally become less distinguishable as the storm moves into northern latitudes, although remnants of them may persist for some time. The winds in storms of tropical origin are characteristically gusty due to instability. Peak wind speeds can typically be 50 per cent greater than the speeds of sustained winds. In combination with topographic effects, wind gusts can be severe and damaging locally (Galbraith 1980; Barks and Richards 1986).

Rainfall

Heavy rainfall is a well-known characteristic of tropical cyclones, with total rainfall amounts in mature hurricanes typically in the 75–250 mm range at the more southern latitudes (Barks and Richards 1986). Maximum rainfall rates usually occur ahead of and to the left of the storm as it undergoes or completes transition. One factor that can reduce the amount of rainfall accumulation in the Maritimes and Newfoundland, compared with that in more southern locations, is the high speed at which many tropical cyclones move through the region. Nevertheless, tropical cyclones can produce extremely heavy rainfalls in the Atlantic provinces on occasion. Hurricane Beth of 1971 brought record rainfall amounts to Nova Scotia. Nearly 250 mm of rain fell in Halifax. Damage to crops, mainly due to flooding, was extensive, and sections of highways and bridges were washed out. Total damage in Nova Scotia was estimated at $3.5 million (CHC 2003). Sometimes tropical cyclones produce relatively light rainfall right after landfall and a torrential downpour a few days later. This occurs when large amounts of tropical atmospheric moisture are released by a passing disturbance or by topographic features such as mountains.

Storm Surge

A *storm surge*, as the name implies, is a surge of water at a coastline that is the direct result of a storm. A storm surge has been defined as a region of elevated sea level generated by strong winds and low atmospheric pressure (McInnes et al. 2003). High alongshore winds 'pile up' the water near the coast, causing the sea level near the coast to rise, and when a hurricane passes, the sea water is further pulled upward in response to the extremely low pressure found near the centre of the storm. The wind ahead of the hurricane plays the key role in setting up large surges at the coastline; sometimes 75 per cent or more of the surge is the result of wind, not low pressure. A storm surge can occur with or without the hurricane making landfall. Surges in the Atlantic are at their highest on the front right side of the hurricane, where alongshore winds are strongest. Various factors affect the intensity of the event. Specifically, the strength of the generating storm, the consistency of wind direction, and the local tidal cycle all play important roles in determining the impact of the surge.

Low-lying areas are especially vulnerable to storm surges, while higher coastal elevations may not be affected (CHC 2003). Globally, storm surges, in combination with high tides, can have devastating consequences. Since 1900, billions of dollars worth of damage to infrastructure and agriculture and hundreds of thousands of deaths have been attributed to them (Heidorn 2005b). In 1869, in New Brunswick and western Nova Scotia, a hurricane known as the Saxby Gale lifted the tide up to 2 m higher than normal. This storm surge, together with a higher-than-average spring tide, had devastating consequences. Many people were caught off guard and lost their lives, hundreds of boats were beached, and all low-lying areas were flooded. Dikes in the vicinity of the Bay of Fundy were breached, and many livestock were reportedly washed out to sea. The quality of the soil on low-lying farmlands suffered for many months from salinization from the floodwaters. Landforms created by the gale still exist today. There are those who believe that the dikes in the Bay of Fundy should be raised so that similar devastation might be prevented in a region whose population is denser now than it was in 1869.

The storm surge created by Hurricane Juan (2003) approached 2.0 m at the head of the Bedford Basin in Halifax Harbour, high enough to erode the coastline, damage boardwalks and infrastructure, cause water to inundate large portions of the Halifax and Dartmouth waterfronts, and flip the railway cars on the train tracks adjacent to the water's edge on their sides. As the hurricane-force winds shifted, the surge quickly dissipated, draining back to the Atlantic and allowing the water in the harbour to return to its original level in an hour and a half (Bowyer 2004). The storm surge from the post-tropical storm Noel damaged beaches along Nova Scotia's Atlantic coast, tossing large rocks onto adjacent roadways and ripping up pavement like paper (*Chronicle-Herald* 2007d).

Modern forecasting equipment has facilitated meteorologists' ability to more accurately forecast storm events and provide timely weather warnings to those in potential danger. As a consequence, the current annual death toll attributed to storm surges is much less than it was in the early and mid-twentieth century (Heidorn 2005b). In their study of storm surge hazards in Canada, Danard et al. (2003) conclude that storm surge risk in Canada is greatest on the Atlantic coast and lowest in the Arctic.

To facilitate the proper assessment of storm surge hazard, they give high priority to the development of maps that would show inundation zones for storms surges that might occur in populated coastal areas.

Tornadoes

Small tornadoes can develop in hurricanes, usually in the front and right side of the storms (in the Northern Hemisphere) as they make landfall and begin to dissipate. The winds at the surface die off quickly, creating a strong vertical wind shear that allows for the development of tornadoes. The damaging winds of hurricanes and tornadoes can sometimes cover the same path, making it difficult to attribute the damage. Tornadoes are more commonly associated with higher-strength hurricanes, such as those that more frequently strike in the Caribbean and the United States. One such example is Hurricane Bertha of 1969, which produced a swarm of over 100 tornadoes on the Texas coast. Typically, the more intense the hurricane is, the greater the tornado threat, and therefore, tornadoes are generally not associated with the tropical cyclones that reach Canada (NOAA 1999; CHC 2003). In the aftermath of Hurricane Juan (2003), however, the impacts of several tornadoes were noted.

The Canadian Hurricane Centre: Forecasting Tropical Storms

The Canadian Hurricane Centre (CHC), a section of Environment Canada's Meteorological Service of Canada branch, has a mandate to conduct research on tropical cyclones and forecast their impact on Canada's land and marine response areas. The CHC forecasts all tropical cyclones that enter the CHC Response Zone, which lies along the Canada–US border and extends 200 nautical miles into Canadian waters (see Figure 6.2 and Table 6.3). From 1997 to 2006, fifty-three tropical cyclones entered the CHC Response Zone, or about 37 per cent of all named Atlantic tropical cyclones in this 10-year period (Bowyer 2007). Another 23 tropical cyclones entered the CHC Response Zone between 2003 and 2006, one of which was the category 2 Hurricane Juan (Figure 6.3). The CHC issues Information Statements when a tropical storm, hurricane, or post-tropical storm is forecast to enter the Response Zone within 72 hours. These statements are issued every six hours until the storm no longer poses a threat to Canadian waters or territory. Position and intensity update bulletins are issued every three hours when tropical storm-force winds (gales) are reported within the CHC's 'area of forecast responsibility'.

The cornerstone of the Canadian Hurricane Centre is a multi-tasking computer workstation that provides meteorologists with simultaneous displays of satellite imagery, numerical guidance, forecast bulletins, and observational data from around the world. CHC forecasters use the Canadian Meteorological Centre's Global Environmental Multiscale (GEM) Model, a dynamic numerical weather model, in addition to a suite of storm-forecasting models produced in the United States and Europe.

> Forecasters at the CHC face many forecast challenges. When a tropical cyclone moves into the mid-latitudes and begins to undergo transition, dynamic models have their largest errors. Further, problems arise when a

Table 6.3 Storms that entered the Canadian Response Zone.
Source: Modified from Environment Canada Centre 2005c.

10-Year Period	Number of Storms	5-Year Period	Number of Storms
1901–10	5	1901–05	3
		1906–10	2
1911–20	4	1911–15	2
		1916–20	2
1921–30	11	1921–25	6
		1926–30	5
1931–40	17	1931–35	8
		1936–40	9
1941–50	13	1941–45	8
		1946–50	5
1951–60	15	1951–55	9
		1956–60	6
1961–70	14	1961–65	9
		1966–70	5
1971–80	9	1971–75	5
		1976–80	4
1981–90	8	1981–85	2
		1986–90	6
1991–2000	15	1991–95	6
		1996–2000	9

tropical cyclone is declared extratropical by the NHC. When this happens, tropical cyclone modeling centers stop modeling the storm. Therefore, forecasters at the CHC lose the majority of their guidance. . . . [U]nfortunately, this may happen at a time when the CHC needs to issue land and marine warnings. (Szymczak and Krishnamurti 2006, 149)

There remain significant challenges in the forecasting of tropical cyclones at extratropical latitudes, and researchers are constantly enhancing models to generate greater lead-times for hurricane and tropical storm warnings in both Canada and the United States. A common problem is the accurate prediction of the behaviour (track, intensity, and impacts) of these rapidly changing systems. The greatest challenges during an extratropical transition event are associated with the potentially large amounts of precipitation, continued high winds, and the generation of large ocean surface waves and swells (Abraham et al. 2004).

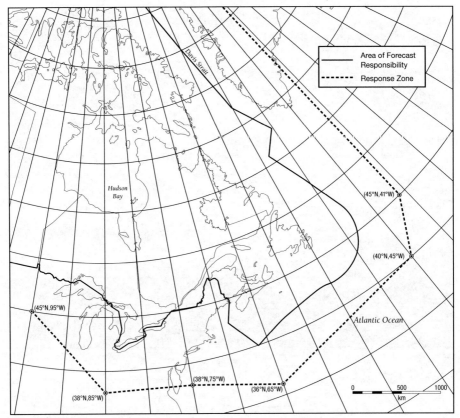

Figure 6.2 Canadian Hurricane Centre area of forecast responsibility and Response Zone.
Source: Modified from Canadian Hurricane Centre 2004.

Tropical cyclone forecasts have greatly improved over the past few decades as a result of weather satellite technology, which can show a storm's precise location and allow tracking at 15-minute intervals. Meteorologists can often give up to five days' warning before a storm strikes land. Forecasters and scientists are now able to collect data from within the storm itself. One tool they use is the *dropsonde*, an instrument that, after being dropped from an airplane, measures the wind speed, air pressure, and humidity within a storm. The data from satellites, dropsondes, and other sources are fed into powerful computers that use mathematical models to simulate the storm's development. Forecasters then study the computer's output to plot the hurricane's expected path. The greatest advances in tropical cyclone forecasting can be attributed to increased computer capacity and speed. New supercomputers now allow for the computation of complex equations on greatly increased resolutions (Bowyer 2007).

Since 1998, synthetic aperture radar (SAR) has been used to scan hurricanes, creating images of areas 450–500 km wide with a resolution where each pixel on the screen is equal to 100 m on the ground. For a hurricane that is not yet over land, the image is a high-resolution picture of the sea surface. The amount of energy on the sea surface reflected back to the satellite, also known as 'backscatter', depends on the roughness of

Figure 6.3 Hurricane Juan approaching Nova Scotia (2003).
Source: Canadian Hurricane Centre. © Her Majesty the Queen in Right of Canada, 2005. Reproduced with the permission of the Minister of Public Works and Government Services Canada.

the surface, which is itself a reflection of wind speed. The faster the winds, the rougher the water, the brighter the image. The calm eye of the hurricane is darker because there are no high winds and the water is calm. Information gathered by RADARSAT (discussed in Chapter 2) is combined with data from optical satellites and from aircraft that fly into hurricanes to track and measure the strength of the storm. The use of both visual and infrared satellite technology enables better forecasting capabilities. The infrared photographs measure the temperature of the surface beneath (land, sea, clouds, etc.) as it is radiated back into space. Visible satellite images provide information about the observed cloud cover as indicated by the amount of solar radiation reflected from the clouds. The pictures allow meteorologists to determine the temperature of the ocean water, the presence of circulation, and the height of moisture in the atmosphere (CHC 2004). (See satellite image of Hurricane Bob in colour insert section.)

The first aircraft reconnaissance undertaken specifically to study a hurricane in extratropical transition was successfully accomplished by the Canadian Convair research aircraft that flew into Hurricane Michael on 19 October 2000. Dropsonde, radar, and cloud microphysical measurements provided a wealth of information on the temperature, wind, and cloud structure near the period of maximum intensity as the hurricane translated rapidly northeastward, south of Newfoundland (Abraham et al. 2004).

In addition to satellite and radar technology, dropsondes, and flight reconnaissance, the Environmental Monitoring Division of the Meteorological Service of Canada,

Atlantic Region, has a series of eight buoys in the Atlantic Ocean that are used in hurricane forecasting. These so-called Environment Canada Oceanographic Data Acquisition Systems (ODAS) weather buoys are moored off the coasts of Nova Scotia and Newfoundland. The data collected from the buoys are used hourly by meteorologists. The buoys measure average wind speed and direction, maximum wind speed, wave height, atmospheric pressure, air temperature, and sea surface temperature. Every hour, the buoys transmit the data via satellite to several weather centres in North America.

Extended-Range Forecasts

There have been a number of attempts in recent decades to produce seasonal extended-range hurricane forecasts. Although the Canadian Hurricane Centre does not presently issue such forecasts, there are a number of examples from the United States and the United Kingdom. Tropical Storm Risk (TSR) in the United Kingdom issues global seasonal forecasts. Colorado State University has been making Atlantic Basin hurricane season forecasts for 24 years. The forecasts are based on statistical methodologies derived from 55 years of data and a study of former seasons that had characteristics similar to those of the current season (Klotzbach and Gray 2007).

The Atlantic Hurricane Seasonal Outlook has been an official forecast product of the NOAA Climate Prediction Center since 1998. Using mathematics, physics, research, global observations, and numerical models, seasonal hurricane forecasters provide the public with

- the probabilities of whether the upcoming Atlantic hurricane season will have an above-normal, near-normal, or below-normal level of activity;
- the likely range of tropical storms, hurricanes, and major hurricanes that can be expected during the season; and
- the scientists' overall confidence level with the forecast.

There are high probabilities for error in the seasonal forecasts, however. In 2006, for example, NOAA initially indicated that the season would have above-normal activity, while in reality the activity was near normal. The conditions that determine active and inactive Atlantic hurricane seasons are largely controlled by recurring rainfall patterns along the Equator. These patterns are linked to two dominant climatic phenomena:

1. The El Niño–Southern Oscillation (ENSO) cycle
2. The tropical multidecadal signal

The multidecadal signal is a major contributor to the observed alternating periods (25–40 years) of active/inactive hurricane seasons. It accounts for the interrelated set of atmospheric and oceanic conditions known to produce active hurricane eras and is strongly related to monsoon rainfall patterns over western Africa and the Amazon Basin and to Atlantic Ocean temperatures. NOAA's seasonal hurricane outlooks are based mainly on the analysis and predictions of the ENSO cycle and the tropical multidecadal signal, along with predictions of upcoming Atlantic Ocean temperatures.

The future of seasonal hurricane prediction will focus on overcoming two major scientific hurdles. The first hurdle is to improve ENSO predictions during what is commonly known as the 'springtime forecast barrier', the period from March through July. Forecasting ENSO during these months can be extremely difficult because the atmosphere is in a state of transition. Yet such forecasts are critical, because they are the only meaningful way to predict the likely strength of storms during the peak of the hurricane season. An uncertain or poor ENSO forecast can lead directly to less confidence in the Atlantic Hurricane Seasonal Outlook (NOAA 2007a). The other major forecast hurdle is to better understand and predict the seasonal factors that control hurricane landfall and then to develop a procedure to make confident hurricane landfall forecasts on seasonal time scales (NOAA 2007a).

Tropical Storms Affecting Canada

Tropical storms show a tendency to follow preferred tracks (Barks and Richards 1986; National Hurricane Center 2008). Three primary offshore tracks have been identified off eastern Canada (Figure 6.4); a minor track through New Brunswick, a second minor track through western Newfoundland, and a third, generally high-storm-

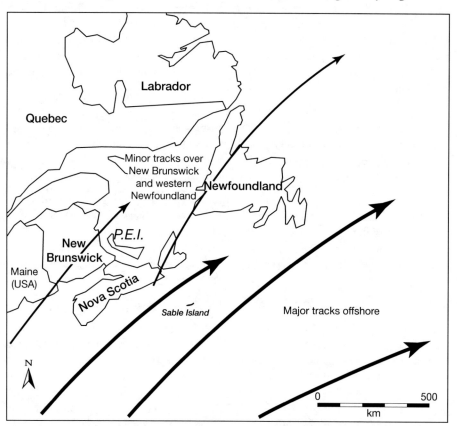

Figure 6.4 Preferred tropical storm tracks in the Northwest Atlantic, 1949–1983.

Source: Modified from Barks and Richards 1986. © Her Majesty the Queen in Right of Canada, 1986. Reproduced with the permission of the Minister of Public Works and Government Services Canada.

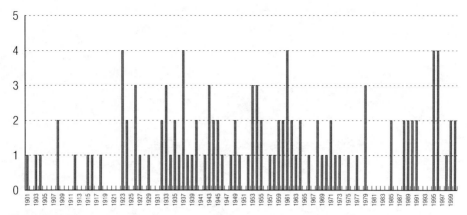

Figure 6.5 Tropical storm frequency by year, 1901–2000.
Source: Modified from Environment Canada 2005c.

frequency track observed south of Nova Scotia. This higher-frequency track may be tied to warmer water temperatures, deeper water, and the natural tendency for storms to dissipate as they move out of the tropics and into the hostile environment of the mid-latitudes. The likelihood of a tropical storm impacting some part of Canada increases significantly towards the latter half of the hurricane season, as ocean waters become warmer at higher latitudes.

On average, about 40 per cent of Atlantic Basin tropical storms threaten to hit Canada each year, although this average masks the large range from year to year. In

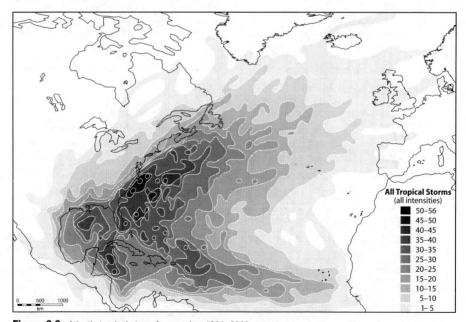

Figure 6.6 Atlantic tropical storm frequencies, 1901–2000.
Source: Modified from Environment Canada 2005c.

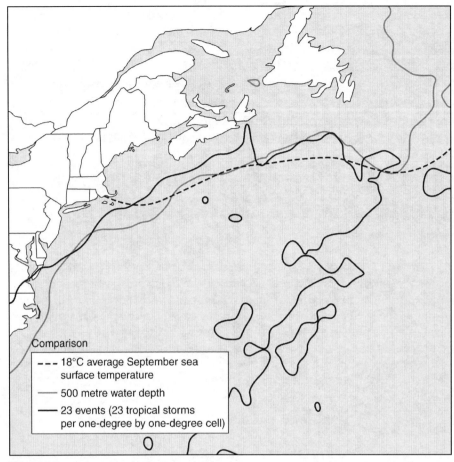

Figure 6.7 Comparison of tropical storm regions (within black line), 18°C average, isotherm, and 500 m water depth, 1949–1983.

Source: Modified from Barks and Richards 1986.

2006, sixty per cent of Atlantic Basin tropical storms threatened some part of Canada, while in 2007, only 13 per cent of the tropical storms had the potential to strike Canada. Eight hundred and seventy-seven storms were reported over the Atlantic Basin between 1901 and 2000, with increased activity in the years 1923, 1937, 1961, 1995, and 1996 (Table 6.3 and Figures 6.5, 6.6, and 6.7) (Environment Canada 2005c). Some of the more notable storms that struck parts of Canada are discussed in some detail in this section.

At least six category 2 hurricanes struck Canada between 1891 and 2003. Although Canada has not been affected by a so-called major hurricane (defined as a category 3 strength hurricane or higher), those that reached Canada have arguably had major impacts. Even though technically category 2 storms, some (such as Hurricane Juan) have caused the kind of destruction more commonly witnessed in the wake of a category 3 storm (Bowyer 2007). Tables 6.2 and 6.4, together with the following chronology, provide statistics and details on the worst storms in Canadian history.

Table 6.4 Notable Canadian hurricanes.
Source: Compiled and modified from PSC 2005a; Environment Canada 2004i, 2005c.

Location	Date	Conditions
Newfoundland	Sept. 1775	The Independence Hurricane hits Newfoundland in September 1775, killing over 4000 people.
Nova Scotia	1863	A category 1 hurricane strikes Nova Scotia, killing 80 people.
Nova Scotia and New Brunswick	1869	The Saxby Gale strikes the Bay of Fundy.
Nova Scotia and Newfoundland	1873	The Nova Scotia Hurricane of 1873 results in 600 deaths.
St Margaret's Bay, NS	1893	Although this storm makes landfall southwest of Halifax, its impact is felt over large areas of NS, NB, PEI, and NL.
Newfoundland	24–25 Aug. 1927	A hurricane sweeps through Atlantic Canada, washing out roads, filling basements, and swamping boats. In Newfoundland, 192 people die at sea.
Southern Ontario	25 Sept. 1941	Winds associated with this storm are reported to reach 130 km/hr in London, ON. Three people are killed.
Charlottetown, PEI	22 Sept. 1942	A storm, remnants of a dissipating hurricane, leaves 163.8 mm of rain and extensive damage in Charlottetown.
New Brunswick	11–13 Sept. 1954	Hurricane Edna causes severe flooding and damage estimated at over $1.5 million. One person is killed.
Toronto, ON	14–15 Oct. 1954	Hurricane Hazel leaves 81 dead and 7472 people (based on 1868 families of 4) homeless; winds reach 124 km/hr; as much as 111 mm of rain fall in 12 hours in some areas; over 210 mm of rain fall over two days, causing severe flooding in the floodplain regions (low-lying areas) of the Don and Humber Rivers and the Etobicoke and Mimico Creeks. The area hit hardest by the storm (in terms of rainfall) is Brampton, and the worst flooding occurs along the Humber River. It is the most severe flooding in the Toronto area in 200 years; 20 bridges are destroyed or damaged beyond repair; the Humber River in Weston rises 6 m, sweeping away a full block of homes on Raymore Drive and killing 32 residents; a trailer park in Woodbridge is flooded and destroyed, killing 20 people. Flooding is widespread, from Lake Simcoe to Toronto and from the Niagara region to Lake St Clair. The hurricane results in over $1 billion in damage.
Newfoundland	Sept. 1958	Although Hurricane Helene was a category 3 storm as it entered Canadian waters, it is quickly downgraded to an extratropical storm when it passes Newfoundland. At 120 km/hr, the winds are strong enough to uproot trees and power lines and to damage homes.

Table 6.4 Continued

Location	Date	Conditions
Maritime provinces	20 June 1959	Hurricane gales, reaching 120 km/hr, sweep across the Maritimes, killing 35 people, mostly lobster fishermen, and causing considerable property damage; 22 fishing boats are destroyed off Escuminac.
Nova Scotia	7–8 Oct. 1962	Hurricane Daisy kills 6 people and causes $10 million (1962 US$) of damage in New England and Nova Scotia. Winds in Nova Scotia gust to 137 km/hr.
Victoria, BC	12 Oct. 1962	Remnants of Typhoon Freda, the infamous 'Columbus Day Storm', strike the Pacific coast; the storm causes 7 deaths; winds recorded in Victoria, BC, reach sustained speeds of 90 km/hr, with gusts to 145 km/hr. Damages are estimated at over $57 million; 20% of Stanley Park is flattened.
Nova Scotia	29 Oct. 1963	Hurricane Ginny passes across southern Nova Scotia with winds up to 176 km/hr. Central Nova Scotia suffers millions of dollars of damage, but there are no fatalities.
Nova Scotia and Newfoundland	21 July 1966	Hurricane Celia strikes southeastern Cape Breton as a category 1 hurricane.
Maritime provinces	21 Oct. 1968	Remnants of Hurricane Gladys pass over Cape Breton Island, killing one person in Nova Scotia and one in PEI and producing rainfall of 45–90 mm; the storm causes flooding in New Brunswick and Nova Scotia.
New Brunswick, Quebec, and Nova Scotia	6 Sept. 1969	Hurricane Gerda strikes St Stephen, NB, and Quebec as a category 2 storm and continues on to strike Labrador as a category 1 storm. This is the only known hurricane to have struck Labrador.
Nova Scotia	15–16 Aug. 1971	Hurricane Beth produces 296 mm of rain in Nova Scotia; damage to crops is extensive; sections of highways and bridges are washed out; freshwater supplies in Dartmouth are left contaminated for days because of extensive runoff into Lake Antigonish. More rain falls during Beth than during Hazel in 1954. One person dies.
Victoriaville, QC	27 Aug. 1972	Hurricane Doria causes floods in the region of Bois-Francs; the communities most affected are Thetford Mines, Victoriaville, and Warwick.
Maritime provinces	27–28 July 1975	Hurricane Blanche brings violent winds up to 130 km/hr and 120 mm of rain, causing major damage.

Table 6.4 Continued

Location	Date	Conditions
New Brunswick and Quebec	10–11 Aug. 1976	Hurricane Belle leaves a rainfall accumulation of 178 mm in a 3-day period, causing extensive damage. There are reports of flooded basements, damage to sewage disposal facilities, and washed-out roads and culverts. On 11 August, the Saint John River rises 4.9 m at Fredericton and 4.3 m at Maugerville. Approximately 13 areas are affected in Quebec, including Notre-Dame-du-Lac and Sherbrooke.
St John's, NL	5 Sept. 1978	A violent storm, Hurricane Ella, with winds over 220 km/hr, passes south of Cape Race; 45 mm of rain and winds of 115 km/h are recorded in St John's.
Southern Quebec	13 Sept. 1979	Hurricane Frederic brings more than 80 mm of rain to southern Quebec.
West coast, BC	11–12 Oct. 1984	A storm unexpectedly forms from the remnants of Typhoon Ogden, causing the death of 5 fishermen.
Nova Scotia	10 Sept. 1989	Hurricane Gabrielle batters the coast of Nova Scotia with 6–9 m swells and winds gusting to 150 km/hr; one man is killed when a huge wave sweeps him into the ocean.
Northumberland Strait, NS, and Prince Edward Island	Aug. 1990	Six crew are killed when a cargo ship south of Nova Scotia is battered by Hurricane Bertha. Winds and heavy rains damage PEI's corn and tobacco crops; huge surf pounds the island's coast; and up to 190 mm of rainfall is recorded in some areas.
Southern New Brunswick and Nova Scotia	19 Aug. 1991	Hurricane Bob hits southern New Brunswick with winds gusting to 100 km/hr after leaving a trail of destruction through the Cape Cod region of the United States. Two people are killed in Nova Scotia.
Burin Peninsula, NL	11 Sept. 1995	Heavy rains and winds associated with Hurricane Luis damage provincial, municipal, and private property on the Burin Peninsula. Some two dozen roads are washed out. One person dies.
Nova Scotia	15 Sept. 1996	Hurricane Hortense makes landfall near Sheet Harbour, NS, as a category 1 hurricane. An estimated $3 million in damage is reported across Nova Scotia.
Newfoundland	19 Oct. 2000	Hurricane Michael makes landfall in Newfoundland as a category 2 storm. This is the first hurricane that Canadian research scientists fly a research plane into. Michael blows roofs off homes, breaks windows, and damages boats.

Table 6.4	Continued	
Location	**Date**	**Conditions**
Southern coast of Newfoundland	19 Sept. 2001	Tropical Storm Gabrielle brings high tides, strong winds, and record-breaking precipitation to the southern coast of Newfoundland, particularly the Avalon Peninsula. Provincial, municipal, and private property is greatly affected. Basements, streets, parks, schools, and shopping malls are flooded. Power outages affect most of the province. Conflict arises between the residents of St John's and the municipality over the inadequate sewer system, which could not manage the runoff. The St John's mayor declares it 'the worst storm in 100 years.'
Cape Breton, Nova Scotia, and Prince Edward Island	2002	Hurricane Gustav makes landfall along the south coast of Cape Breton in the early hours of 12 September 2002. It delivers over 100 mm of rain in parts of Nova Scotia and winds of 100 km/hr or more throughout large areas of Atlantic Canada. Damage includes localized coastal flooding from a storm surge in the southern Gulf of St Lawrence and downed trees from high winds across Prince Edward Island. There is one death in PEI.
Halifax, NS, and Charlottetown, PEI	29 Sept. 2003	At 12:10 a.m. ADT, Hurricane Juan makes landfall in Nova Scotia as one of most powerful and damaging hurricanes to ever hit Canada. The category 2 hurricane packs winds reaching up to 174 km/hr. There are 8 deaths.

1775: The 'Independence Hurricane' passed Newfoundland on 9 September 1775, killing over 4000 people, many of them seamen from Britain and Ireland. Despite the existence of first-hand accounts of the storm, it may never be possible to know the true death toll (Ruffman 1995). It is known that a great number of fishing boats were out in Conception Bay, as the squid catch was late that summer. The men were oblivious to the growing winds and the rapid approach of the storm. The sea is said to have risen 6 m higher than usual, putting the boats in the bay at great risk. After the weather system had moved on and died down, the beaches were littered with the corpses of the dead sailors. It is said that for many years afterwards bones were washed ashore and that, according to local settlers, the waves held the 'sound of the desperate cries from the drowning men' (BBC 2006).

Newfoundland's fisheries 'received a very severe stroke from the violence of a storm of wind, which almost swept everything before it,' the colonial governor Robert Duff wrote shortly after the Independence Hurricane struck. Most of the reports of hurricanes before 1775 are lost to history. It is largely because Duff wrote a report to his superiors in England that we know so much about the 'Great Newfoundland Storm', as it has also been called. The Independence was Atlantic Canada's first recorded hurricane and one of Canada's most tragic natural disasters. It is also rated as the eighth deadliest Atlantic hurricane in history (Wikipedia 2007c).

1869: The 'Saxby Gale' struck New Brunswick and Nova Scotia on 4–5 October 1869. The gale was named after Lieutenant Saxby of the Royal Navy. A year earlier, he had predicted that a severe storm accompanied by a very high tide would hit somewhere on the planet. Saxby knew that on 5 October 1869 the moon would make its monthly passage closest to Earth (perigee) and that there would be a new moon on the same day (Environment Canada 2007c). Many people in the United Kingdom, Canada, Newfoundland, and the United States dismissed Saxby's predictions, thinking them lunatic, as they were accustomed to frequent gales and hurricanes during the month of October. On the afternoon of 4 October in Saint John, New Brunswick, the wind increased to gale force. Rain began to fall at 6:00 p.m. By 8:30 p.m., the wind was blowing at hurricane force, reaching its maximum velocity at about 9:00 p.m. By 10:00 p.m. the wind began to subside and shifted to the southwest. Most of the storm's damage was due to an enormous storm surge. The Saxby Gale storm surge produced a water level that gave Burntcoat Head, Nova Scotia, the honour of having the highest tidal range ever recorded, and in Moncton the tide reportedly rose nearly 2 m above former records. The storm produced waves that, with the storm surge, breached the dikes that served to protect the low-lying farmland, sending ocean water far inland. People and farm animals drowned in the floods that followed, and hundreds of boats were beached. Over 100 people were killed in the storm (Environment Canada 2007b). Compared to New Englanders, who paid no attention to British news, many in the Canadian Maritimes were aware of the forecast and did prepare, ultimately mitigating what would have been a greater loss of life.

1873: The 'Nova Scotia Hurricane of 1873', also known as the 'Great Nova Scotia Cyclone', caused heavy damage and loss of life in the month of August. The hurricane paralleled the Nova Scotia coastline offshore, bringing heavy winds and rain to the province. It lost intensity and had weakened to category 1 by 25 August while south of Newfoundland. It swept across Cape Breton, slowed, drifted northward, and then struck Newfoundland, making landfall on 26 August.

Despite its relatively low maximum winds, the Great Nova Scotia Cyclone was a deadly storm. In Nova Scotia, it destroyed 1200 boats and 900 buildings, and killed 500 people, mostly sailors who were lost at sea. Damages amounted to $3.5 million (1873 US$, or $53.9 million in 2005 dollars). Its impact was also felt in Newfoundland, where the storm's winds and floods killed an additional 100 people. Losses were high in part because the telegraph service between Toronto and Halifax was interrupted, preventing storm warnings from getting through. Of significance to Canadian meteorology, this storm, perhaps more than any other event, convinced officials of the need for an improved Canadian storm-warning system (Environment Canada 2007c).

1893: The '1893 Hurricane', sometimes referred to as the 'Second Great August Gale' (the Nova Scotia Hurricane of 1873 being considered the first), affected large areas of Nova Scotia, Prince Edward Island, New Brunswick, and Newfoundland. It made landfall just west of Halifax in St Margaret's Bay as a category 1 hurricane (65 knots). This hurricane landed in the midst of a very active hurricane season, and the weather systems in the days and weeks prior to and after this event were particularly hard on both Atlantic Canada and the New England states. These systems wreaked havoc with the shipping and fishing industries, with many vessels sunk, lost at sea, or suffering great damage. Newspapers were full of reports of the losses at sea, often written in the typical prose of the time: 'The storm king had no pity, and his agents, the remorseless waves, cruelly did his bidding and swallowed up the helpless victims, twenty-four in number' (*Halifax Herald,* 25 August 1893).

The damage that the 1893 Hurricane caused in Halifax was very similar to that inflicted by Hurricane Juan a hundred years later. Winds reached 96 km/hr in the city, knocking down trees, damaging infrastructure, overturning trains, and knocking out power. Newspaper editorials of the day criticized the power company and suggested that electrical lines should be buried underground—a suggestion that continues to be made. One individual noted that with the advent of electricity, storms now brought a new fear—the fear of the dark when the power goes out.

1900: 'The Galveston Hurricane' of September 1900 took its name from the widespread disaster it wrought on the coastal Texas town of Galveston. The storm began to form on 27 August 1900 off the Cape Verde Islands in the Atlantic Ocean. After its landfall in Texas, it tracked northwards through the American Midwest, entered the CHC Response Zone on 11 September, and hit Canada on 12 September east of Port Albert, Ontario. The storm crossed southern Ontario, the Gaspé Peninsula of Quebec, New Brunswick, and Newfoundland before exiting the CHC Response Zone on 14 September. Maximum wind speeds for this storm were 213 km/hr (115 knots).

The Galveston Hurricane was responsible for as many as 250 deaths in Canada, mostly due to the loss of fishing and shipping vessels off Newfoundland and Prince Edward Island, in addition to unconfirmed deaths from St Pierre et Miquelon. Crop damage amounting to an estimated $1 million was experienced in Ontario. New Brunswick, Prince Edward Island, and Nova Scotia sustained wind and rain damage to crops and structures (Environment Canada 2005c).

In the period between the late 1920s and the 1950s, most people who perished as a result of tropical storms and hurricanes lost their lives on vessels at sea. There weren't significant land-based impacts from storm systems on land.

1927: The '1927 Nova Scotia Hurricane' (also known as the '1927 Great August Gale' or the 'Great Gale of August 24') was a powerful category 2 hurricane that struck Nova Scotia in mid-August. It made landfall northwest of Yarmouth, Nova Scotia, and crossed the province with winds reaching 166 km/hr. It made landfall in eastern Prince Edward Island and western Newfoundland early the next day with the same intensity it had had when it first struck Nova Scotia. As many as 192 people lost their lives in this storm, most at sea (Environment Canada 2005c). Ships ahead of the hurricane received repeated storm-warning broadcasts from both US and Canadian weather officials. Small-craft warnings and a hurricane warning for New York City were also issued. Unfortunately, the majority of fishing vessels in Atlantic Canada in this period did not have radios, leaving the large fishing fleet on the offshore banks unaware of the approaching disaster. Later, the extratropical remnants of the hurricane were tracked as far north as Iceland.

Nova Scotia received up to 102 mm of heavy rainfall and experienced gale-force winds. Flooding washed out 20–25 per cent of the rail lines across the province, disrupting rail service, and further inhibited travel by making numerous roads impassable and sweeping away bridges. The storm inflicted severe crop damage, destroying 50 per cent of the fruit, vegetable, and hay harvest. The 1927 hurricane had followed the tragic August storm of the previous year, and together they became known as the 'August Gales', remembered for generations as being among the worst tragedies in the fisheries of Maritime Canada. The heavy shipping losses, especially among fishing schooners on the banks, accelerated a move to outfit Canadian schooners with motors and radios. Today the 1927 Nova Scotia Hurricane is commemorated in a waterfront monument and an exhibit at the Fisheries Museum of the Atlantic in Lunenburg, Nova Scotia.

1950: 'Hurricane Able', the first Atlantic tropical cyclone in the era of named storms, made landfall on the morning of 21 August just west of Halifax as a tropical storm with winds of 65 km/hr (35 knots). Able subsequently made landfall over Newfoundland as a tropical depression with winds of 55 km/hr (30 knots), on 22 August.

1954: 'Hurricane Hazel' of 1954 is Canada's most-remembered 'hurricane'. In actuality, it wasn't a hurricane when it hit Canada, but rather a dying tropical cyclone that re-intensified rapidly. Up to 200 mm of rain fell in less than 24 hours near Toronto, resulting in 81 deaths and over $100 million of damage (Environment Canada 2004g). Flooding from Hazel was disastrous in parts of southern Ontario, with estimates of as much as 90 per cent of the precipitation running directly into rivers, raising the water level by 6–8 m (Environment Canada 2005c). Houses were torn from their foundations, and cars and mobile homes were entrained in the floodwaters. The Weather Office was simultaneously praised for its accuracy and emphatic warnings and denounced for not providing enough warning. The rapid flooding caught many area residents by surprise. Despite repeated warnings about the amount of rain expected, they were unable to imagine how that might affect them, having never experienced floods of Hazel's magnitude. Many people were drowned trying to escape from their homes, were washed off roads, or were swept down the river still in their houses. The estimated cost of Hazel was $100 million (approximately $1 billion in 2005 Canadian dollars).

Significant outcomes of Hazel were the formation of the Toronto and Region Conservation Authority (TRCA) and the prioritization of flood control and flood warnings by the three levels of government. Conservation authorities were granted the power to buy and regulate floodplain land, and in cooperation with the three levels of government, they helped create flood-control and flood-warning systems. Floodplains were bought from private property owners, and subsequent development was restricted on the lands.

1959: The 'Escuminac Hurricane' (or 'Escuminac Disaster'—Escuminac is a community located on the eastern side of New Brunswick, facing the Northumberland Strait) was not formally named during the 1959 season, although modern analysis indicates that it did reach hurricane strength before it became extratropical. On 20 June, this rare early hurricane system made landfall as a category 1 hurricane at Canso, Nova Scotia, and moved over the Northumberland Strait with winds of 130 km/hr (70 knots). The storm made landfall in Prince Edward Island as a tropical storm with winds of 111 km/hr (60 knots) later that day. It claimed the lives of 35 men, most of them fishermen from Prince Edward Island and New Brunswick. As in the 1927 Hurricane, the fishermen who perished were not aware of the storm warnings, since most boats at the time had no radios. Of the 54 boats that had sailed from Escuminac that day, 22 were lost.

Her Majesty Queen Elizabeth II and His Royal Highness Prince Philip, Duke of Edinburgh, on a royal tour in another part of Canada at the time, expressed their sympathies. They were reportedly the source of a large anonymous donation that was made to the relief fund in the days following the event. Today, the Escuminac Disaster Monument sits as a memorial on the shores of Escuminac Harbour, not far from the wharf from which the fleet sailed.

1963: 'Hurricane Ginny' made landfall as a category 1 hurricane at Yarmouth, Nova Scotia, on 29 October 1963, with winds of 176 km/hr (95 knots). Ginny then passed over Moncton, New Brunswick, and finally over northern Prince Edward Island, with category 2 hurricane winds of 166 km/hr (90 knots). There would not be another category 2 hurricane to make landfall in Nova Scotia for 40 years (when Hurricane Juan arrived in 2003) (Environment Canada 2005c). Hurricane Ginny is unique because at landfall in Nova Scotia in late October it was still a hurricane. Historically, it remains the latest landfalling hurricane to affect the New England states and Canada. Ginny remains one of the strongest landfalling hurricanes on record for Nova Scotia, and yet the newspapers of the time reported that although Ginny was a hurricane when it hit the province, Nova Scotia 'escaped the brunt of the storm' (*Chronicle-Herald* 1963).

From the late 1970s though the 1980s, there was a relative lull in hurricane activity. Although some notable tropical systems moved through eastern Canada in the latter part of the twentieth century, these did not have far-reaching impacts; nor did they significantly affect large population centres in the way that Hurricane Juan would in 2003. It is therefore not surprising that Hurricane Juan came as such a surprise to most people.

1991: Referred to as the 'Halloween Storm' or the 'Perfect Storm', this unnamed storm made landfall over Nova Scotia and Prince Edward Island on 2 November as a tropical storm with winds of 93 km/hr (50 knots). This complex storm was not easy to forecast. It began with a strong extratropical low that formed off the coast of Nova Scotia on 28 October. The low moved southward, developing into an extratropical storm south of Halifax. This was the first storm in which the moored-buoy program off Canada's east coast recorded phenomenal wave heights, including a single wave of over 30 m (Environment Canada 2005c). The storm was made infamous with the publication of Sebastian Junger's book *The Perfect Storm*.

1996: Hurricane Hortense struck eastern Nova Scotia on 15 September 1996. This category 1 hurricane attained winds of 130 km/hr (70 knots) and was the first hurricane to make landfall in Nova Scotia since 1975. It resulted in many power outages, trees blown down, roofs torn away, and roads damaged. An estimated $3 million of damage was reported for Nova Scotia alone. Significant wave heights were recorded, with maximum waves near 18.0 m. Highest rainfall amounts exceeded 135 mm, and large amounts fell in short periods of time (18–22 mm per hour in some areas). In some cases, the amounts falling within one or two hours exceeded normal monthly totals (Environment Canada 2005c). Some Haligonians may remember this storm, not for its high winds and heavy rain, but for the fact that the power went off merely minutes before the beginning of the final game in the World Cup of Hockey between Canada and United States.

2000: Hurricane Michael made landfall in Newfoundland near Harbour Breton on 19 October, crossed the province, and then turned east before dissipating at sea on 20 October. Maximum sustained wind speeds associated with this category 2 hurricane were 158 km/hr (85 knots), noted within the CHC Response Zone and at landfall. There were numerous reports of damage from across Newfoundland. Most of the damage, while light, was due to high winds and was reported from small communities east of where landfall had occurred. Hurricane Michael gave Canadian scientists from Environment Canada and the National Research Council their first opportunity to fly a research plane into a tropical cyclone (Environment Canada 2005c).

2003: Hurricane Juan made landfall just west of Halifax on 29 September as a category 2 hurricane with 158 km/hr (85 knot) winds (Figure 6.8). Less than three hours later, it made landfall in Prince Edward Island as a category 1 hurricane. The storm entered the CHC Response Zone on the evening of 27 September as a category 2 hurricane and remained at that strength at landfall. The strongest one-minute sustained winds reported were 158 km/hr (85 knots) at McNabs Island in Halifax Harbour. The highest peak gust reported was 232 km/hr (125 knots), from the vessel *Earl Grey* while at anchorage in the Bedford Basin at the head of Halifax Harbour.

Rainfall with Juan was quite low, with the highest official accumulation of 38 mm at Halifax International Airport. The storm surge near and just east of Halifax was estimated at close to 2 m, however, establishing a new record water level at Halifax. A significant wave height of 11.5 m (the average of the highest one-third of the waves) and a maximum wave height of 19.9 m were calculated at the mouth of Halifax Harbour (Environment Canada 2005c; Bowyer 2008). Shortly before the storm made landfall, the mayor of Halifax declared a state of emergency. This was well warranted, given that roofs of buildings were collapsing or were seriously damaged, including the roof of a local hospital. Trees across the city fell, some striking vehicles, and train cars were derailed. Live explosives that had harmlessly lain on the bottom of Halifax Harbour since the World War II were swept ashore in Bedford Basin during the storm and had to be detonated by the navy the following week.

Juan is believed to have been the most widely destructive tropical cyclone to hit Atlantic Canada in over a century, with an estimated loss of 100 million trees in Nova Scotia and one million trees in Halifax alone. In the landmark Point Pleasant Park, which was hit particularly hard, an average of five trees fell per second during the three-hour storm. Ninety per cent of the mature growth in the park was destroyed or irreparably damaged. Juan claimed eight lives, four directly (two inland and two

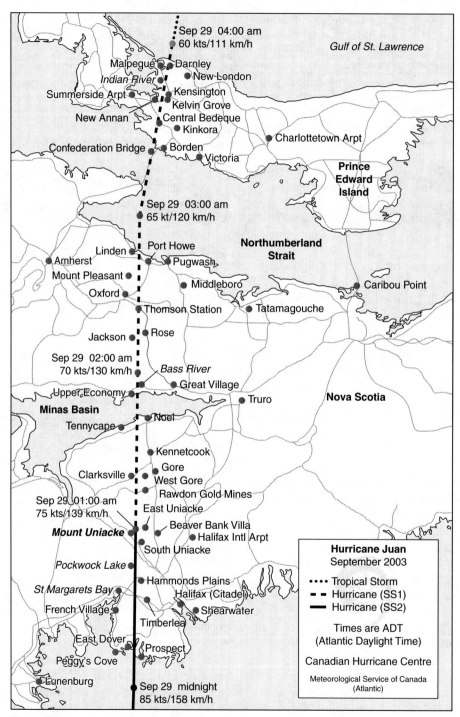

Figure 6.8 The path of Hurricane Juan, 2003.

Source: Modified from Environment Canada 2005c. © Her Majesty the Queen in Right of Canada, 2003. Reproduced with the permission of the Minister of Public Works and Government Services Canada.

Figure 6.9 Destruction caused by Hurricane Juan.
Source: Cathy Conrad.

marine) and another four in the aftermath. Power outages in Nova Scotia and Prince Edward Island left over 300,000 customers without power for up to two weeks. Nearly all commercial activity ceased in the Halifax area for two to five days, and schools were closed for a week. Dozens of marinas in Nova Scotia and Prince Edward Island were destroyed, and dozens of small vessels capsized or sank (Figure 6.9). Public reports of 'a wall of water' moving onto the coast and up Halifax harbour coincided with the arrival of the highest surge and waves. Estimates of $100 to $150 million damage are conservative. The name 'Juan' was retired from the World Meteorological Organization's list of hurricane names (Environment Canada 2007f).

Among the thousands of individual stories is one about the Sambro Island lighthouse (the oldest lighthouse in North America), which sustained damage to the base of the tower and to the gas house from the storm surge. In Stewiacke, Truro, Tatamagouche, and Pictou County, barns and silos were flattened or damaged. Forestry operations in the centre of the province were greatly affected by 'wind bombs' that completely flattened sections of forest on various woodlots. Following the onset of Hurricane Juan, a vast number of previously rarely seen birds began to turn up on Nova Scotia shorelines (Environment Canada 2007f).

Less than one year before Juan struck Atlantic Canada, a study of hurricane risk and perception in Nova Scotia was undertaken (Hanson 2002). This study determined that although the majority of people believed that the chances were fair to good that a hurricane could strike Nova Scotia, they tended to think it unlikely that damage would result. Given that the general public made light of a hurricane's possible effect (most never having lived through one), that the forecasters under-predicted the event

('Juan is expected to make landfall as a weak hurricane', *Chronicle-Herald* 2003), and that local Emergency Management officials were unprepared and caught off guard, it was fortunate that Juan's impact was not greater. It helped that the hurricane struck just past midnight, when most people were at home. Four years later, the city of Halifax was still 'dealing with the aftershocks' of Juan (*Chronicle-Herald* 2007a).

Climate Change and Tropical Cyclones

Atmospheric scientists are in universal agreement that the Atlantic Basin has returned to an extended period of heightened hurricane activity. In fact, the years 1995–2005 saw the highest frequency of tropical cyclones of any decade on record. And yet debate remains concerning the specific links between climate change and hurricane frequency and intensity. Climate science in relation to climate change and increased hurricane frequency and intensity is still in its infancy. Some researchers do not support a strong linkage (e.g., Pielke et al. 2005; Trenberth 2005), while others (e.g., Emanuel 2005; Anthes et al. 2006) are of the opposite opinion. Since 1995, the number and intensity of hurricanes in the Atlantic have increased, but it has been argued that this can be attributed to multidecadal variability, which has been well documented since the 1900s (Gray et al. 1997; Landsea et al. 1999; Goldenberg et al. 2001). However, the documentation of trends in hurricane intensity in relationship to increasing sea surface temperatures does suggest that the increasing intensity of hurricanes may be attributed to global warming.

One problem in the analysis of data in the field of hurricane climatology is that the inconsistency in existing hurricane records may make the accurate measurement of trends difficult. Scientists are attempting to construct more homogeneous global records of hurricane intensity (Kossin et al. 2007). Hurricane activity comprises three factors:

- Number of storms in a season (frequency)
- Duration of storms (duration for an individual storm is defined as the period of time in which its intensity, measured as a maximum sustained surface wind speed, is greater than 17 m/s–1)
- Intensity of storms

The relationship between hurricane activity and sea surface temperature (SST) can be viewed as part of a larger relationship between hurricane activity and Atlantic variability, referred to as the Atlantic Meridional Mode (AMM). The AMM is correlated with a number of local climatic factors (SST, shear, low-level vorticity and convergence, static stability, and sea-level pressure) that cooperate to increase or decrease Atlantic hurricane activity (Vimont and Kossin 2007).

> The persistently warm phase of the AMM that the Atlantic is experiencing at present, and the significant upward trend of the AMM since 1970 supports the idea that hurricane activity is in a long-term active phase. But it is unclear at present whether this will continue indefinitely or whether hurricane activity will return to a more quiescent period similar to the 1970s and 1980s.

Much of the uncertainty lies in the larger uncertainties regarding how anthropogenically forced tropical Atlantic SST increases will project onto future hurricane variability. (Vimont and Kossin 2007, 12)

Anthes et al. (2006) note that several studies (Landsea et al. 1998; Pielke et al. 2005) have concluded that the recently observed increases in Atlantic hurricane frequency and intensity are within the range of observed multidecadal variability. A number of multi-year atmospheric and oceanic patterns of variability, including the El Niño–Southern Oscillation phenomenon, the quasi biennial oscillation, the North Atlantic Oscillation (NAO), and the Atlantic Multidecadal Oscillation (AMO), have had a strong impact on hurricane frequency, intensity, and tracks. But it is not valid to therefore assume that all the variability during the twentieth century has been natural. Tompkins (2002) goes so far as to say that it would be 'reckless' not to expect that climate change will have some impact on extreme weather events. According to Tompkins, while there are obvious large and natural oscillations, the growing body of evidence suggests a direct and growing trend in several important aspects of tropical cyclones, such as intensity, rainfall, and sea level, all of which can be attributed to global warming. Pielke et al. (2005) assert, however, that linkages between global warming and hurricane impacts are premature for three reasons:

1. No connection has been made between greenhouse gas emissions and the observed behaviour of hurricanes (Houghton et al. 2001; Walsh 2004).
2. The peer-reviewed literature suggests that any future changes in hurricane intensities will likely be small in the context of observed variability (Knutson and Tuleya 2004, Henderson-Sellers et al. 1998).
3. Anticipated future storm damage to society is likely to be more the result of growing wealth and population than of the behaviour of hurricanes (Pielke et al. 2000).

With respect to the first reason, Emanuel (2005) has made further suggestions about linkages and is continuing to conduct research in this area.

While tropical storms do seem to follow certain natural cycles, scientists are nevertheless concerned about the effects of global warming and long-term climate change. On 15 September 1999, the United Nations issued a report predicting that global warming will cause more frequent and more severe tropical storms, floods, and tornadoes in the coming century. Hurricane intensity is directly correlated with the warmth of ocean waters. Global warming is going to continue, and this will result in warmer ocean temperatures. We can therefore expect to see the signal of stronger hurricanes. Scientists still have to contend with the fact that the effects of human activities are superimposed on the various natural cycles also known to be present. Regardless of attribution (natural variability, anthropogenic forcing, or some combination of the two), there has been a remarkable increase in Atlantic hurricane activity since 1970 (Kossin and Vimont 2007).

In November 2006, one hundred and twenty-five leading tropical cyclone researchers from 34 different countries and regions met at the World Meteorological Organization's International Workshop on Tropical Cyclones (IWTC). At the conclusion of the invitation-only workshop, held once every four years, the participants issued a statement comprised of twenty-seven key points. The final point reads as follows: 'Despite the diversity of research opinions on this issue it is agreed that if there has been a recent increase in tropical cyclone activity that is largely anthropogenic in origin, then humanity is faced with a substantial and unanticipated threat' (IWTC 2006, 7). It is against this backdrop that Canadians must consider meaningful adaptation and preparedness mechanisms to address the impacts of the potentially more severe hurricanes of the future.

Chapter Summary

Although hurricanes (also referred to as tropical cyclones) often bypass eastern and western Canada offshore or strike with less intensity than elsewhere, it is important to realize that hurricanes and tropical storms do occur in Canada every year. Weather maps illustrate that we are affected by hurricanes more often than most of us realize, although Canada is usually only struck by weaker storms because of the cooler sea surface temperatures immediately offshore. There are no records of a category 3, 4, or 5 hurricane ever making landfall in Canada, although an average of four tropical cyclones hit Atlantic Canada every year. Statistically, Newfoundland has the highest probability of being struck by a tropical cyclone. A transitioning cyclone can have devastating effects on Canada even without a high Saffir Simpson classification. Not only will a larger area be hit by hurricane- or gale-force winds, but the heaviest precipitation can be directly over land, even if the extratropical cyclone does not make landfall. Approximately 46 per cent of the tropical cyclones in the Atlantic Ocean undergo extratropical transition. Impacts from tropical storms include heavy rains, high winds, storm surge, and the potential for tornadic activity. Climate science in relation to climate change and increased hurricane frequency and intensity is still in its infancy. Researchers are divided on whether or not there are strong linkages.

Chapter Seven

Canadians and Weather: Vulnerability, Risk, Adaptation, and Mitigation

> **Objectives**
>
> - To review the concepts of vulnerability, risk, adaptation, and mitigation in the context of severe Canadian weather
> - To relate these concepts to the implications for climate change

Introduction

We have reviewed how the various forms of severe weather affect us. In what ways are we vulnerable? How might we better respond to, mitigate, or adapt to such extremes? On 5 December 2006, the *Abbotsford News* reported that a storm that had struck British Columbia's Lower Mainland in November of that year 'could have been a human catastrophe for Abbotsford's homeless, as record snowfall, plunging temperatures and bitter wind made spending a night outdoors a potentially fatal proposition.' The community, fortunately, had a plan for extreme weather and was able to put mitigation strategies in place to prevent such a catastrophe.

Society in general has gradually begun to move towards the adoption of mitigation strategies in its approach to severe weather so as to avoid impacts that otherwise might result in natural disasters. This shift in thinking, however, is only slowly emerging, as the general approach is still overwhelmed by attitudes that favour response measures rather than proactive tactics. For some time to come, it is likely that mitigation efforts will exist in the shadow of well-established response efforts. Strongly influencing the degree of interest that citizens, response organizations, and politicians will have in mitigation strategies are the disasters that will challenge them to reassess their perception of risks and the actions they might take to reduce losses (Newton 2002). There remain plenty of issues to debate; vulnerability, risk, hazard, resilience, and resistance are just some of the concepts widely used in disaster management for which there are still no universally agreed-upon definitions (Bogardi 2006).

Vulnerability

A society's vulnerability is broadly understood as its predisposition to be hurt should an event beyond a certain (though ill-defined) threshold of magnitude occur; this 'hurt' could be inflicted on its population, its economic assets, its infrastructure, or the ecosystem (Bogardi 2006).

If hazards remain constant (in frequency and severity) while disaster losses (damages caused by the events) increase, vulnerability (exposure to damage) must be to blame. Although we are unable to stop or control most types of hazards, we can certainly reduce, though not completely eliminate, their effects—that is, our vulnerability to them (McEntire 2005). Some practitioners (e.g., Bogardi 2006) believe that vulnerability assessment will become a crucial component of disaster preparedness. In monitoring vulnerability, we can identify those target communities where proactive measures are needed, and thus we can pre-empt the most devastating consequences of extreme events should they occur. We could greatly improve our ability to assess a population's vulnerability (and then apply this information in the policy- and decision-making sphere) if we were to develop indicators or indices to encapsulate the notion of vulnerability (Bogardi 2006). Globally, those members of society who are particularly vulnerable to increased threats from severe weather and climate change are those who face poverty, unequal access to resources, food insecurity, conflict, and a high incidence of diseases such as HIV/AIDS.

Over the past few decades, the challenge of reducing socio-economic vulnerability to climate and weather-related hazards has been taken on by four distinct research and policy communities, their areas of focus being disaster-risk reduction, climate-change adaptation, environmental management, and poverty reduction (Thomalla et al. 2006). While the concepts of *risk* and *vulnerability* may be linked in their relation to hazards, they differ profoundly from one another: risks are taken, while vulnerability merely exists. People have the choice to undertake risky behaviour (e.g., going to the beach to see the high waves when a storm surge is forecast versus heeding warnings and staying indoors), but they may have little or no control over their vulnerability (e.g., not having access to information or the means of evacuation in the event of a natural disaster). In other words, risks are voluntarily entered into, or at least can often be avoided, while vulnerability must be endured (Alexander 1997). Risk is the product of hazard (the physical agent and its impact), and vulnerability is the susceptibility to damage or injury. The roots of vulnerability to local disasters are increasingly recognized to be the pre-existing patterns of community settlement and development. The damage to any particular household is the result of a complex set of interacting conditions, some having to do with geography and location, some with the dwelling, and still others with the social and economic characteristics of the people living there (Hearn Morrow 1999).

Alexander (1997) identifies different forms of vulnerability:
- Economic vulnerability (loss of occupation or income through the destruction of the means of production or commerce)
- Technological vulnerability (of the rich; this can involve heavy losses from a natural disaster in absolute terms, but is more often an inconvenience than a personal tragedy)

- Newly generated vulnerability (refers to risks to property or other capital assets that are newly constructed in hazard zones, or to new residents who lack protection against hazards with which they are unfamiliar)
- Residential vulnerability (includes pre-code buildings, un-reinforced historical structures, and other sources of risk that have not been modified to modern safety standards)
- Delinquent vulnerability (occurs when safety norms, codes, and regulations are flouted or ignored)

After post-tropical storm Noel struck Nova Scotia in November 2007, with wind gusts up to 135 km/hr, a well-known folk art carver in East Ship Harbour said that the storm had wrecked his livelihood (*Chronicle-Herald* 2007d). It had lifted his painting shed about 3 m and rolled it into the woods, destroying several pieces of art as well as all of his art supplies, illustrating this individual's economic and residential vulnerability to a storm of this magnitude. The predominant view of vulnerability takes into account the needs of women, children, minorities, the elderly, and people with disabilities, as these populations often require special assistance before, during, and after disasters (McEntire 2005). But we might benefit from an even broader conceptualization of vulnerability. It cannot be denied, for example, that the wealthy and privileged are more vulnerable where high-end development has taken place in hazard-prone areas.

King (2000) describes three facets of community vulnerability: poor understanding and use of the extensive information available on hazardous events; people's isolation during hazardous events; and a general lack of adequate awareness and preparedness. Bankoff (2003) expands on this, identifying six categories of vulnerability that depend on the characteristics of the individual, organization, or entity affected: physical, personal, structural, institutional, economic, and cultural vulnerability.

Meteorologist Peter Bowyer, a former director of the Canadian Hurricane Centre, experienced the aftermath of Hurricane Juan and subsequently reflected on the conditions that he believes increase the vulnerability of society. A slide presentation of his ideas, entitled 'Vulnerability and Hurricanes', includes the following insights:

- *Low expectations lead to higher vulnerability.* People who had previously experienced a hurricane were more likely to expect impacts, although their expectation was still that impacts would be low (Hanson 2003).
- *Low awareness leads to higher vulnerability.* Many people remain unaware of emergency management preparedness measures, for example (Hanson 2003).
- *Denial leads to higher vulnerability.* Individuals sometimes simply don't believe that something bad could happen to them.
- *Ignorance leads to higher vulnerability.* The general public, as well as decision-makers, should be aware that different weather patterns, different weather elements, and different time/space scales differ in terms of predictability.
- *'Ignore-ance' leads to higher vulnerability.* A particularly serious problem arises when severe events fail to follow forecasts of storms (the 'cry wolf' syndrome), as this puts the public at risk of becoming complacent when future warnings are issued.
- *'Boring-ness' leads to higher vulnerability.* Certainties are more readily accepted by the public than uncertainties. Certainties are news, uncertainties aren't.

- *Miscommunication leads to higher vulnerability.* A high level of media attention to a potential storm, coupled with experience (or lack of) of similar recent events, can contribute to people's sense that this is a deterministic, rather than a probabilistic, event; the uncertainty of the event needs to be more adequately discussed. Broadcast meteorologists and weathercasters work under significant competitive pressures, leading some of them to express unrealistic confidence in their predictions.

In a 2005 study of risk and perceptions associated with hurricanes in Nova Scotia, respondents were nearly equally split between those who felt vulnerable to hurricanes (53 per cent) and those who did not (47 per cent). The respondents in the former group cited the following reasons for their sense of vulnerability:

- Personal experience (30 per cent)
- The physical characteristics of hurricanes (24 per cent)
- The location of their homes (20 per cent)
- The possible failure of technology (12 per cent)
- Climate change (12 per cent)

The question remains whether people's awareness or understanding of their *potential* vulnerability in the event of a hurricane indirectly reduces their risk of being affected by such an event.

Resilience

The concept of resilience is also linked to vulnerability. Resilience describes the capability of a system to maintain its basic functions and structures in a time of shocks and perturbations (Birkmann 2006). The resilience perspective implies that we can respond and recover (McEntire 2005). Resilience is distinct from resistance (ability to mitigate disaster). Geis (2000) argues that we should be focusing on 'disaster resistant' communities as opposed to 'disaster resilient' communities. Given that resilience is the ability to recover from or adjust easily to misfortune or change, while resistance is merely the ability to resist, we must ask ourselves whether we want our communities to 'recover from or adjust easily to a disaster' (we are assuming that one has occurred), or whether we want them to 'resist the disaster' (i.e., not allow the inevitable damage from an extreme natural event to reach 'disastrous' proportions) (Geis 2000, 152).

The World Conference on Disaster Reduction (WCDR) held in Kobe, Japan, in January 2005 formulated the goal of creating societies that are more resilient in the event of disaster. Key to the accomplishment of this goal would be the development of a system of indicators of disaster risk and vulnerability. Such a system would enable decision-makers to assess the potential impact of disasters and to promote the formulation of appropriate policy responses while identifying the most threatened areas and social groups (Birkmann 2006). The WCDR conference led to the creation of a blueprint, agreed to by participating governments, called the 'Hyogo Framework for Action 2005–2015: Building the Resilience of Nations and Communities to Disasters'.

Perception of Risk

'Just like the characters in animated films who, suspended in mid-air, do not plunge to the ground until they realize their predicament, people construct their own reality and evaluate risks according to their own subjective perceptions.'

Jaeger et al. 2002

One of the most serious obstacles for emergency measures planners in any region is the general public's lack of awareness. A person's perception of the environment and of the risks inherent to certain aspects of the environment describes that person's level of awareness. People's level of awareness may be disproportionate to the actual risk as interpreted by scientists and government advisors (Bruce et al. 2000; Etkin 1999). The behaviour of individuals before, during, and after hazardous events has been the focus of many studies by researchers from a range of academic disciplines, including geography, sociology, anthropology, psychology, and medicine.

Risk perception is an individual's intuitive impression of the probability of a hazard affecting them (Leiss and Chociolko 1994). Researchers (e.g., Jones and Andrey 1998) have recognized that a person's experience with a hazard is a major factor in his or her perception of risk, and the perception of risk will determine how a person responds to warnings or forecasts. Risk entails more than the probability of an event, although that is often how it is measured. Any analysis of risk must include vulnerability, which would take into account, among other things, absolute and relative measures of the population and property at risk (Tobin and Montz 1997).

$$\text{Risk} = \text{probability of occurrence} \times \text{vulnerability}$$

A sample survey of Nova Scotians prior to Hurricane Juan indicated that the majority of people underestimated the risk of a hurricane strike. Although 63 per cent thought it was somewhat likely that a hurricane would affect the area where they lived, only 38 per cent felt they might suffer from personal or property damage as a result of such an event. Only 13 per cent of those surveyed were aware of municipal preparations for an event, and even fewer were aware of provincial or federal preparations (Conrad and Hanson 2004).

The 'risk triangle' attributed to Crichton (1999) illustrates that risk is the probability of there being harmful consequences or losses as a result of exposure to a given hazard (a given element of danger or peril) over a specified time period. The area of the triangle represents the risk, while the sides of the triangle represent the hazard (e.g., flood water), exposure (housing on a floodplain), and vulnerability (the housing has been flood-proofed). Altering any side of the triangle alters the risk.

Individuals' perceptions of risks are relevant to the policy process. The degree of risk individuals assign to activities involving possible harm helps shape their attitudes towards public policy on such issues. As the perception of future hazard severity increases, support for government intervention to address the hazard will increase (Gerber and Neeley 2005). Tobin and Montz (1997) caution that although we may be tempted to assume that individual perceptions will lead to actions, after or in anticipation of an event, directed towards reducing individual and community vulnerability, frequently such actions do not occur. Individuals may feel relatively powerless when faced with the extremes of the natural environment and with the realities of major political, social, and economic structures that constrain individual activity. In addition, the actions of some individuals may not be considered appropriate by others. As Tobin and Montz (1997) note, 'Faced with expected flooding, some individuals may sandbag their homes while others pray for divine intervention to stop the floodwaters. Others, when warned that mobile homes are inappropriate housing within a coastal flood or hurricane zone, may respond that mobile homes are cheaper and quicker to replace than more permanent structures' (133).

Tobin and Montz (1997) pose three questions to illustrate the range of factors that contribute to what some might view as irrational behaviour in the face of a hazardous event. First, why do so many individuals relocate to areas of known problems? Second, why do many individuals risk their lives by remaining in their homes in spite of repeated warnings at the onset of an event? Third, why do individuals prior to and during hazardous events exhibit behaviour that can appear bizarre? The explanation for behaviour lies in both cognitive and situational factors. *Cognitive* factors include personality characteristics that influence one's view of nature and hence one's propensity to avoid or take a risk. *Situational* factors complicate an individual's range of choices; they include a person's physical location in relation to a hazardous area as well as income, age, and social factors that may affect an individual's ability to undertake specific actions.

A perception that hurricanes, earthquakes, tornadoes, and other hazards are not serious threats would be part of a person's cognitive set. Why would anyone take remedial action if he or she did not believe that there was a high risk of the event occurring? The same attitudes operate for hazard-warning systems. Unless a threat is perceived as imminent, the warning may be ignored. If previous warnings resulted in 'unnecessary' evacuation because nothing untoward occurred, residents might decide to sit the next one out (the 'cry wolf' syndrome). Warnings that are not followed by a hazardous event can lead to unsatisfactory behaviour in subsequent events. Individuals may become blasé about warnings and indifferent to the hazard, and consequently they may take little or no remedial action (Tobin and Montz 1997).

Uncertainty is a complication that must be recognized and addressed by those charged with managing risk. The public derives most of its knowledge of risk from weather experts and the media. However, experts tend either to over- or to underestimate the public's ability to evaluate risk and choice. Whichever the case, the result is the same: appropriate information is not communicated to the public in an understandable form (Tobin and Montz 1997). To communicate risk appropriately, the conveyors of risk information must find out what people already know, must develop warning messages on that basis, and, finally, must test the messages for successful communication.

Research undertaken by the National Science Foundation in the United States has tackled the question of why people tend to live in places with risk in the first instance. This research has shown that, in many cases, emotional benefits interfere with an individual's ability to assess risk. People who live in risk-prone areas may focus on the things they love about their location, such as its environmental beauty or proximity to the ocean, and simultaneously discount the risk. Researchers have found a correspondence between perceived risk and perceived benefit in people's emotional evaluations of a potential hazard. If people like an activity, they judge the risks as low. If they dislike an activity, they judge the risks as high. For example, people buy houses or cars that they like and that appeal to them emotionally, and they downplay the risks associated with the purchase. This may explain why people sometimes make seemingly irrational, high-risk decisions, such as settling along the coastline where there is greater vulnerability to earthquakes and hurricanes. Or when they travel, they may choose to explore beautiful but dangerous spots. People also make decisions about risk on the basis of their own feelings about the available information on a hazard. Facts alone are often not enough to change people's perceptions of risk. Individuals need to relate to those facts on an emotional level for their risk judgment to be affected.

Care needs to be taken in the crafting of 'vivid messages' or 'fear appeals' as well. Some vivid messages can lead to undesirable responses. Teenagers might actually be more inclined to take up smoking, for example, in response to certain fear appeals. Etkin and Myers (2000) have observed that, despite the effectiveness of the media in communicating to the public, media images over time may serve to desensitize people to risks.

According to Jacqueline Meszaros of the National Science Foundation (CBC Radio One 2007), people have an internal risk-reward ratio, asking themselves: Is the risk too great for the reward? Meszaros suggests that they tend to differentiate between categories of risk, being more worried about unfamiliar, unnatural risks. Given that people are relatively familiar with natural hazards (e.g., fire or weather), they worry less about them than about unfamiliar hazards (e.g., technological or terrorist threats). This type of intuitive risk perception is based on how information on the hazard (the source of risk) is communicated, on the psychological mechanism people use to process uncertainty, and on people's earlier experience of danger. This mental process results in perceived risk, which can be summed up as being a collection of notions that people form about risk sources based on both the information available to them and their basic common sense (Jaeger et al. 2002).

Some researchers have spent time evaluating the different ways to view risk. Renn (2004) differentiates between our perceptions of 'risk as a fatal threat', 'risk as a personal thrill', and 'risk as a game of chance'. In the context of risk as fate, natural disasters are usually seen as unavoidable events that have catastrophic results, but they are also seen as quirks of nature or acts of God and thus beyond human control. People are not yet sufficiently aware of the possibilities for controlling natural disasters and lessening their impacts to allow the risks associated with them to be assessed in the same way as those from technological accidents (e.g., risks associated with nuclear power plants). Natural burdens and risks are seen as an almost inevitable prescribed fate, while technical risks are seen as the consequences of decisions and actions. The rarer the event, the more likely people are to deny or suppress the idea of its occurring; the more frequent the event, the more likely it is that people will flee from the danger zone. As opposed to the case with technical risk, the random nature of the natural disaster is not the fear-triggering factor (because randomness involves fate and not the unforeseeable consequences of inappropriate action).

Probability and severity of adverse effects are not the only factors that people use as yardsticks for perceiving and evaluating risk. Research has provided a lengthy list of supporting circumstances or qualitative factors that influence an individual's ability to perceive and evaluate risk:

- Familiarity with the risk source
- Voluntary acceptance of the risk
- Ability to control the degree of risk personally
- Ability of the risk source to cause a disaster (catastrophic potential)
- Certainty of fatal impact should the risk occur (dread)
- Undesired impact on future generations
- Sensory perception of danger
- Impression of reversibility of the risk impact
- Reliability of information sources
- Clarity of information on risk

Studies conducted on an international scale also show that people everywhere, regardless of their social or cultural background, use much the same risk-assessment criteria in forming their opinions.

Mitigation and Adaptation

Although mitigation and adaptation are two different approaches to climate change and severe weather, they are often incorrectly used interchangeably. In the most fundamental sense, *mitigation* is a proactive measure to lessen the impact of an event. *Adaptation*, however, carries the implication that the inevitable impact will occur but that adaptation strategies will be in place to deal with it.

Adaptation can be viewed as an individual's or a community's (or an ecosystem's) ability to adjust to an actual or expected stimulus or its impacts. It is also viewed as a conscious ongoing process of monitoring, evaluating, and learning to make decisions. These decisions might involve changes to processes, practices, or structures in order to reduce potential vulnerabilities and damages or to take advantage of new opportunities that may emerge. Climate adaptation depends on the adaptive capacity of a system, whether it be an ecosystem, a human settlement, or a country. A country's adaptive capacity depends on its infrastructure, wealth, institutions, access to resources, the education and skills of its population, and consequently its vulnerability (Cohen et al. 2004). The United Nations Development Program (UNDP) identifies three forms of adaptation:

- *Autonomous adaptation:* An unconscious process of system-wide coping, most commonly understood in terms of ecosystem adjustments
- *Reactive adaptation:* A deliberate response to a climatic shock or impact, the purpose being to recover and to prevent similar impacts in the future
- *Anticipatory adaptation:* A planned action to prepare for and minimize impacts (e.g., of climate change)

Within the community involved in climate-change adaptation, it is commonly asserted that an improvement in our ability to predict the magnitude and frequency of severe events will enable us to provide more effective adaptation strategies

(Thomalla et al. 2006). For this reason, the community emphasizes the importance of developing hazard forecasting and early warning systems. An alternative view is that if we could cope better with present climatic risks (possibly through improved current forecasts), we could significantly reduce the impacts of future climate change. These views have not been comprehensively tested, however, and it remains unclear how improved forecasting can be used in reducing social vulnerability (Thomalla et al. 2006). The UNDP is presently urging less-developed countries to design National Adaptation Programs of Actions (NAPAs) to address their urgent climate-change adaptation needs. A vast array of adaptive responses are available to nations and communities, ranging from technological (e.g., sea defences), through behavioural (e.g., altered food and recreational choices) and managerial (e.g., altered farm practices), to planning (e.g., planning regulations) measures (IPCC 2007a). The Intergovernmental Panel on Climate Change (IPCC) offers the following caution in its 2007 report:

> While most technologies and strategies are known and developed in some countries, the assessed literature does not indicate how effective various options are at fully reducing risks, particularly at high levels of warming and related impacts, and for vulnerable groups. In addition, there are formidable environmental, economic, informational, social, attitudinal and behavioural barriers to the implementation of adaptation. . . . Adaptation alone is not expected to cope with all the projected effects of climate change. (IPCC 2007a, 19)

The definition of mitigation adopted through a Canadian symposium on the topic provides a context for practical recommendations and actions: hazard mitigation is sustained action to reduce risk to life, property, and the environment from hazards (Greene 1998). Newton (2002) defines mitigation as measures taken in advance of a disaster to reduce the impact (loss of life, damage to property, infrastructure, etc.) of that disaster on the society immediately at risk. Mitigation options for climate change are ever decreasing, and adaptation may increasingly be the only option, although there remain significant options for the mitigation of severe weather. These include everything from seeding clouds (to inhibit hail) to purchasing personal insurance. In general, efforts to mitigate disasters are understood to be positive actions undertaken to reduce the severity, violence, cruelty, or pain of prevailing or future conditions (Newton 1997).

Mitigation actions can be a blend of policies, educational programs, physical structures (such as dams), the development of resistant or resilient systems, retrofitting, or land use planning. Such actions affect both the social and the natural realms. Generally, mitigation occurs through activities that (1) reduce risk and/or (2) transfer or share risk. Risk reduction can be accomplished by modifying the hazard or reducing vulnerability. Studies of some hazard-reduction programs, such as weather modification, either have had mixed results or have not been encouraging. In fact, the American Meteorological Society policy statement on planned and inadvertent weather modification says that 'there is no sound physical hypothesis for the modification of hurricanes, tornadoes, or damaging winds in general' (WMO 1995; NOAA 2003). Other strategies, such as floodways, dikes, land use planning, re-vegetation of slopes, and irrigation, can be very effective and have been widely used (Etkin and Leman Stefanovic 2005). The Red River Floodway, built in the 1960s to protect the city of Winnipeg from flooding in the Red River Basin, is one of the best-known examples of investment in disaster mitigation in Canada (see Chapter 5). Disaster mitigation strategies in place in Canada include the following:

- Hazard mapping
- Adoption and enforcement of land use and zoning practices
- Implementing earthquake-resistant building codes
- Enforcing fire-resistant building codes
- Floodplain mapping
- Reinforced tornado-safe rooms
- Burying of electrical cables to prevent ice buildup
- Dike building and the raising of homes in flood-prone areas
- Disaster-mitigation public-awareness programs
- Insurance programs
- Hail suppression

Hearn Morrow (1999) argues that effective hazard mitigation and emergency response must begin with the understanding that the social, economic, and political structures of our society tend to foster important differences in the vulnerability of those they are meant to protect and serve. This tendency can only be countered by community planning at the local level, where even the most disenfranchised stakeholders are included in the process (Hearn Morrow 1999). Disaster mitigation must stress social rather than physical approaches; such social approaches must place an emphasis on proactive rather than reactive actions; and proactive actions need to focus on internal flaws in society rather than on external forces.

The following principles underlie the Canadian Institute for Catastrophic Loss Reduction's approach to mitigation (ICLR 1998):

- The threat of severe weather is increasing, but sustained action can reduce catastrophic losses. Hazard assessment and risk identification are cornerstones of catastrophic-loss mitigation.
- Solid applied research provides an essential foundation for effective action to reduce future losses.
- Those who knowingly choose to assume greater risk must accept an increased degree of responsibility for their choice.
- Communication with the public before a peril strikes is an important means of reducing losses.
- Local and individual actions are the most effective means of reducing the loss of life and property.
- Partnership is the best approach to resolving shared problems, particularly public safety concerns.

The Context for Canada

Since 1996, Canada has experienced a significant increase in the costs incurred by the Disaster Financial Assistance Arrangement (DFAA), the primary mechanism by which the Government of Canada provides assistance to Canadians affected by natural disasters. The physical devastation and economic losses resulting from the Saguenay River flood (1996), the Red River flood (1997), and the eastern Canadian ice storm (1998) exposed the susceptibility of Canadians to major natural hazards. Together, these events affected approximately 20 per cent of the Canadian population and cost the Canadian government an average of $366 million each in disaster financial assistance payments. In contrast, prior to 1996, the Canadian government's disaster assistance costs per incident had not exceeded $30 million.

In Canada, as in other parts of the world, the tendency towards more disasters and escalating disaster costs seems inevitable. Compared with other countries, Canada should be well positioned to adapt to climate change, given its relative wealth, high literacy rate, and its well-developed infrastructure and resource management systems. Despite these advantages, however, recent years have seen an increase in damage costs from climatic events (Cohen et al. 2004). Processes such as urbanization, globalization, climate change, and reliance on technologically based and interdependent infrastructure can significantly increase risks and damage costs, both direct and indirect; they will also add to the complexity of disaster management in the future, which will include the establishment of an efficient national emergency management system that encompasses mitigation, preparedness, response, and recovery (Hwacha 2005).

Dore (2003) estimated that Canadians can anticipate at least 12 hydro-meteorological disasters annually, with costs estimated up to $1.8 billion. Curtailing this escalating trend requires focusing on reducing disaster vulnerability and protecting Canada's economic and social assets through concerted efforts in disaster mitigation. Both public and private sectors are now considering how they might better mitigate their losses in the future by creating more disaster-aware and disaster-resilient communities (Cohen et al. 2004). In Canada, the insurance industry works towards the goal of mitigation mainly through the research and lobbying efforts of the Institute for Catastrophic Loss Reduction (www.iclr.org), which held a series of workshops across Canada in 1998 with the purpose of developing a national mitigation strategy (ICLR 1998).

The structure of Canada's emergency management system is shaped by its legislative, regulatory, and policy frameworks. The Emergency Preparedness Act (1988) outlines the emergency preparedness role and responsibilities of federal departments and establishes the federal government's relationship with provincial and territorial governments, which in turn delegate responsibility to local-level authorities. Canada's current emergency management approach remains overtly response-focused. Recurrent natural disasters, anticipated increases in hydro-meteorological disasters due to climate variability, and potential disaster-related costs to society are pressuring all levels of government to modernize the existing emergency management system (Hwacha 2005).

The Canadian Risk and Hazards Network, the Institute for Catastrophic Loss Reduction, Public Safety Canada, the Geological Survey of Canada, Environment Canada, and the Canadian Centre for Emergency Preparedness all take on some mitigation functions and, given the mandate and additional resources, have the potential to assume much larger roles (Etkin and Leman Stefanovic 2005). The development of emergency planning and response activities in Canada had its initial impetus in the formation of a civil defence organization in 1948 as part of Canada's participation in NATO. Until the mid-1960s, Canadian civil defence agencies focused their attention on the threat of nuclear explosions. By 1966, the need for a coordinated federal response to peacetime disaster was recognized, and the responsibility was assigned to the Emergency Measures Organization (Newton 1997).

Today, mitigation receives comparatively less attention than preparedness, response, or recovery, making it the least-developed component in Canada's emergency management system, although there have been efforts to address this. Public Safety Canada is currently developing Canada's National Disaster Mitigation Strategy (NDMS). The goals of the strategy are to reduce the risks, impacts, and costs associated with natural disasters, as well as to foster a disaster-resilient society. The strategy aims to encompass structural and non-structural mitigation. Structural mitigation includes such measures as construction of floodways and dikes and the reinforcement of structures against floods, fire, earthquakes, and tornadoes. Non-structural measures include hazard mapping and risk assessments, insurance and public awareness programs, and building code enforcement, standards, and legislation (http://www.publicsafety.gc.ca/prg/em/ndms/index-eng.aspx).

The Canadian government is working towards what is referred to as an 'all-hazards approach' to disaster mitigation—that is, an approach entailing all potential risks and impacts, natural and human-induced (intentional and non-intentional)—to ensure that decisions made to mitigate against one type of risk do not increase our vulnerability to other risks. On its website, Public Safety Canada encourages individual citizens to take precautions in advance of severe weather or other high-risk events and offers advice to families and individuals in the event of severe storms.

Immunity from the effects of extreme natural phenomena is unrealistic for any country, especially one with the geographic scope of Canada. However, given the uneven distribution of the Canadian population, the potential for significant losses is concentrated within a small portion of the country. Moreover, a climate of extremes has conditioned Canadians to adapt to a variety of severe weather conditions more readily than residents of countries with more moderate climates (Newton 1997).

There remain particularly dramatic and pressing concerns for the people in Canada's northern communities, where the impacts of increasingly dire climate scenarios are already being felt. The predominant hazards for those living in isolated northern communities are flooding, forest fires, and severe winter storms. The resilience of Canada's northern inhabitants in the face of such natural hazards is remarkable; it underscores the adaptive capacity developed over generations of experience with the vagaries of the natural environment. However, only a few of the research studies published to date have considered the more immediate natural hazards now affecting northerners and indigenous people (Newton et al. 2005). Increased dependence on southern-style community services, changes in social structures, and the passing away of the traditional life on the land have caused a slow erosion of this historical coping capacity. And this is happening when natural hazard events are becoming more frequent, are increasing in intensity, and are expanding in range (Newton et al. 2005).

Conclusion

At all levels of government, comprehensive hazards mitigation strategies have been moderated to focus on the extension, not the expansion, of services. To date, Canada has been very fortunate compared to many countries, and yet it is becoming increasingly vulnerable (Newton 1997). The recent concept of sustainable hazards mitigation (incorporating a long-term perspective through improved engineering, safer urban development, and cautious environmental management) might provide some direction for the future, but there are many critics of this concept. Perhaps scholarship and policy around emergency management are currently floundering because we are still failing to understand and accept responsibility for the vulnerability factor in disasters (McEntire 2005).

More recently, however, adaptation and vulnerability to climate change, rather than strictly severe weather, have become the subject of considerable attention. The Fourth Assessment Report of the IPCC (2007a, Working Group II) reviews the current knowledge about responses to climate change, including adaptation and vulnerability, and concludes that there have been increasing efforts through the early part of the twenty-first century to adapt to observed and anticipated climate change. The report cites such examples as the coastal defence projects in the Maldives and The Netherlands and the Confederation Bridge in Canada. Other examples include measures to prevent glacial outburst flooding in Nepal and water management strategies in Australia. The IPCC report also alludes to the European countries' plans to address future heat waves. The work of the IPCC underscores the fact that the world now faces severe and hazardous impacts for which adaptation is the only available and appropriate response; in other words, some adverse impacts from climate change can no longer be mitigated. And beyond this, the IPCC report suggested that unmitigated climate change would 'in the long term, be likely to exceed the capacity of natural, managed and human systems to adapt' (IPCC 2007a, 20).

A variety of adaptation options are available to nations and communities, but the IPCC advises that, to reduce vulnerability to the impacts of future climate change, more extensive adaptation than is currently being practised is required. The same can be said for severe weather, since, as this text has outlined, there are strong linkages between many kinds of severe weather and climate change. With the goal of decreasing the vulnerability of the population in mind, we would do well to use a combination of adaptation and mitigation strategies in scenarios involving both severe weather and climate change.

It has become increasingly apparent that we need to develop better adaptation strategies to deal with the forces in society that make us vulnerable to weather and climate (Cohen et al. 2004). The damage and loss incurred as a result of numerous severe weather events in Canada—examples of which have been presented throughout this text—underscore this fact. Given the evidence that our climate is changing, and with it the occurrence and distribution of severe events, there is a growing and increasingly urgent need that we review and expand our present mitigation and adaptation strategies.

References

Abraham, J., J.W. Strapp, C. Fogarty, and M. Wolde. 2004. 'Extratropical transition of Hurricane Michael'. *Bulletin of the American Meteorological Society* 85 (9): 1323–39.

Agriculture and Agri-Food Canada. 2007. 'Drought Watch'. www.agr.gc.ca/pfra/drought/index_e.htm.

Aguado, E., and J. Burt. 2007. *Understanding weather and climate*. 4th ed. Toronto: Pearson Education.

Akinremi, O., and S. McGinn, 2001. 'Seasonal and spatial patterns of rainfall trends on the Canadian Prairies'. *Journal of Climate* 14:2177–82.

Akinremi, O., S. McGinn, and H. Cutforth. 1999. 'Precipitation trends on the Canadian Prairies'. *Journal of Climate* 12:2996–3003.

Alexander, D. 1997. 'The study of natural disasters, 1977–1997: Some reflections on a changing field of knowledge'. *Disasters* 21 (4): 284–304.

ALTAWATCH. 2006. www.raes.ab.ca/abwatch.html.

AMS (American Meteorological Society). 2008. 'Glossary of meteorology'.http://amsglossary.allenpress.com/glossary.

Angas, K. 2006. 'In the last ten years, what natural disasters has Canada experienced? What was the cost to repair the damages? Where does Canada get its aid money?' *EnviroZine*, Environment Canada's online newsmagazine, vol. 34. www.ec.gc.ca/envirozine/english/issues/34/print_version_e.cfm?page+questions.

Annan, K. 2004. Preface to *Living with risk: A global review of disaster reduction initiatives*. International Strategy for Disaster Reduction. www.e11th-hour.org/public/natural/living.with.risk.html.

Anthes, R.A., R.W. Corell, G. Holland, J.W. Hurrell, M.C. MacCracken, and K.E. Trenberth. 2006. 'Hurricanes and global warming—Potential linkages and consequences'. *Bulletin of the American Meteorological Society* 87 (5): 623–8.

Ashmore, P., and M. Church. 2001. 'The impact of climate change on rivers and river processes in Canada'. *Geological Survey of Canada Bulletin* 555.

Atmospheric Environment Service Drought Study Group. 1986. *An applied climatology of drought in the Prairie provinces*. Canadian Climate Centre Report No. 86–4. Downsview, ON: Atmospheric Environment Service.

Bailey, W.G., T.R. Oke, and W.R. Rouse. 1997. *The surface climates of Canada*. Montreal and Kingston: McGill-Queen's University Press.

Bains, N., and J. Hooey. 1998. 'Before lightning strikes'. *Canadian Medical Association Journal* 159 (2): 163.

Bankoff, G. 2003. 'Constructing vulnerability: The historical, natural and social generation of flooding in metropolitan Manila'. *Disasters* 27 (3): 224–38.

Barks, E.A., and W.G. Richards. 1986. 'The climatology of tropical cyclones in Atlantic Canada'. Internal Report # MAES 4-86, April. Atmospheric Environment Service, Bedford, NS.

Bass, B., E. Krayenhoof, A. Martilli, R. Stull, and H. Auld. 2003. 'The impacts of green roofs on Toronto's urban heat island'. In *Proceedings of the First North American Green Roof Conference: Greening rooftops for sustainable communities,* 292–304. Chicago, IL: Cardinal Group.

BBC (British Broadcasting Corporation). 2006. 'Great storms: Hurricane 1775'. www.bbc.co.uk/weather/features/storms_hurricane2.shtml.

Belmin, J., J-C. Auffray, C. Berbezier, P. Boirin, S. Mercier, B. de Reviers, and J-L. Golmard. 2007. 'Level of dependency: A simple marker associated with mortality during the 2003 heat-wave among French dependent elderly people living in the community or in institutions'. *Age and Ageing* 36 (3): 298–303.

Birkmann, J., ed. 2006. *Measuring vulnerability to natural hazards.* Tokyo: United Nations University Press.

Bogardi, J. 2006. Introduction to *Measuring Vulnerability to Natural Hazards,* ed. J. Birkmann, 1–6. Tokyo: United Nations University Press.

Bonsal, B., and R. Lawford. 1999. 'Teleconnections between El Niño and La Niña events and summer extended dry spells on the Canadian Prairies'. *International Journal of Climatology* 19:1445–58.

Bonsal, B.R., X. Zhang, L.A. Vincent, and W.D. Hogg. 2001. 'Characteristics of daily and extreme temperatures over Canada'. *Journal of Climate* 14:1959–76.

Bonsal, B., G. Koshida, E.G. O'Brien, and E. Wheaton. 2004. 'Droughts'. In *Threats to water availability in Canada.* NRWI Scientific Assessment Report Series, No. 3.

Bowyer, P. 1995. *Where the wind blows.* St John's, NL: Breakwater Books.

——— 2004. 'The storm surge and waves at Halifax with Hurricane Juan'. www.atl.ec.gc.ca/weather/hurricane/juan/storm_surge_e.html.

——— 2007. Meteorologist, Environment Canada's Canadian Hurricane Centre. Personal communication.

——— 2008. Meteorologist, Environment Canada's Canadian Hurricane Centre. Personal communication, 8 May.

Brandon, E.W. 1983. 'Atlantic tropical cyclone tracks'. Internal Report # MAES 2-83, February. Atmospheric Environment Service, Bedford, NS.

Brazel, A.J., and R. Osborne. 1976. 'Observations of atmospheric thermal radiation at Windsor, Ontario, Canada'. *Archiv für Meteorologie, Geophysik und Bioklimatologie* B24:189–200.

Brooks, G.R., S.G. Evans, and J.J. Clague. 2001. 'Flooding'. In *A synthesis of natural geological hazards in Canada,* ed. G.R. Brooks, 101–43. Ottawa: Geological Survey of Canada Bulletin 548.

Bruce, J., I. Burton, and M. Egener. 'Disaster mitigation and preparedness in a changing climate: A synthesis paper'. Emergency Preparedness Canada.

Brun, S.E., D. Etkin, D. Gesink Law, L. Wallace, and R. White. 1997. 'Coping with natural hazards in Canada: Scientific, government and insurance industry perspectives'. Study written for the Round Table on Environmental Risk, Natural Hazards and the Insurance Industry.

Bullock, T. 2007. Science and Technology and Transfer Manager, Environment Canada. Personal communication.

Burrows, W.R., and R.A. Treidl. 1979. 'The southern Ontario blizzard of January 26 and 27, 1978'. *Atmosphere-Ocean* 17 (4): 306–20.

Burrows, W.R., P. King, P.J. Lewis, B. Kotchtubajda, B. Snyder, and V. Turcotte. 2002. 'Lightning occurrence patterns over Canada and adjacent United States from Lightning Detection Network observations'. *Atmosphere-Ocean* 40 (10): 59–81.

Burt, C. 2004. *Extreme weather: A guide and record book.* New York: W.W. Norton & Company.

Canadian Climate Impacts Scenarios Project. 1999–2004. www.cics.uvic.ca/scenarios/bcp/select.cgi?&sn=2.

Canadian Disaster Database. 2007. www.publicsafety.gc.ca/res/em/cdd/index-eng.aspx.

Canadian Encyclopedia, The. 2007. 'Cold Places in Canada'. www.thecanadianencyclopedia.com/index.cfm?PgNm=TCE&Params=A1ARTA0010200.

Canadian Geographic. 2007. 'Floods'. www.canadiangeographic.ca/SpecialFeatures/Floods/sagfact.asp.

CHC (Canadian Hurricane Centre). 2002a. 'Hurricane Gustav storm summary'. www.atl.ec.gc.ca/weather/hurricane/gustav02_e.html.

——— 2002b. '2001 Tropical Cyclone Season Summary'.

Canadian Institute for Climate Studies. 2001. Information www.cics.uvic.ca/climate/index.htm.

Canadian Red Cross. 2006. '2005 western Canada floods and severe weather response'. www.redcross.ca/article.asp?id=01383&tid=081.

Cannon, T. 1994. 'Vulnerability analysis and explanation of "natural" disasters'. In *Disaster, development and environment*, ed. A. Varley, 13–30. London: John Wiley and Sons.

Carter, A.O., M.E. Millson, and D.E. Allen. 1989. 'Epidemiological study of deaths and injuries due to tornadoes'. *American Journal of Epidemiology* 130 (6): 1209–18.

Catto, N., D. Ingram, E. Edinger, D. Foote, D. Kearney, G. Lines, and B. Whiffen. 2006. 'Impacts of storms and winds on transportation in southwestern Newfoundland'. Paper presented at the Annual Meeting of the Canadian Association of Geographers, Thunder Bay, ON, Program 2006, 34–5.

CBC. 2003. 'Environment Canada may close weather offices'. 9 January. www.cbc.ca/canada/story/2003/01/09/envirocan030109.html

——— 2004a. 'Fog killed songbirds in Bay of Fundy'. www.cbc.ca/news/story/2004/06/03/fog_birds040603.html.

——— 2004b. '1930s Drought'. www.cbc.ca/news/background/agriculture/drought1930s.html.

——— 2005. 'Storm warnings: Too little, too late?' 2 August.

——— 2006a. 'B.C. hot and getting hotter'. 21 July. www.cbc.ca/canada/british-columbia/story/2006/07/21/bc-heat.html.

——— 2006b. '"Everything's gone," says survivor of deadly tornado in Manitoba'. 7 August. www.cbc.ca/canada/manitoba/story/2006/08/07/man-tornado-aftermath.html.

——— 2006c. 'Heat wave continues across Ontario'. 2 August. www.cbc.ca/canada/ottawa/story/2006/08/02/heat-alert.html.

——— 2006d. 'Historic' October snowstorm blasts Niagara Region'. 13 October. www.cbc.ca/canada/toronto/story/2006/10/13/snow- niagara.html.

——— 2006e. 'Humidity keeps Halifax firefighters busy'. 21 July. www.cbc.ca/canada/nova-scotia/story/2006/07/21/humidity-alarms.html.

——— 2006f. 'Lightning strikes seniors home, leaving 85 homeless'. 1 August. www.cbc.ca/canada/montreal/story/2006/08/01/seniors- tues.html.

——— 2006g. 'Manitoba seeks better weather warning system'. 8 August.

——— 2006h. 'Ontario's sizzling heat breaks records'. 2 August. www.cbc.ca/canada/toronto/story/2006/08/02/heat-ontario-records.html.

——— 2006i. 'Snow leaves 15,000 in B.C. without power'. 28 October. www.cbc.ca/canada/story/2006/10/28/bc-powercuts.html.

——— 2006j. 'Snowstorm blankets B.C.'s southern coast'. 26 November. www.cbc.ca/canada/story/2006/11/26/vancouver-snow.html.

——— 2006k. 2 killed in Quebec storm, thousands without power. 2 August. www.cbc.ca/canada/montreal/story/2006/08/02/quebec-storm.html.

——— 2006l. 'Visitors scramble as water shortage shuts Tofino businesses'. 30 August. www.cbc.ca/canada/british- columbia.story/2006/08/30/tofino-water.html.

——— 2006m. 'Weather warnings too slow: Expert'. 7 June.

——— 2006n. 'Winnipeg sets new records for hot, dry weather'. 1 August. www.cbc.ca/canada/manitoba/story/2006/08/01/july-weather.html.

——— 2007a. 'Forces of nature'. www.cbc.ca/news/background/forcesofnature/gfx/tornado1.gif.

——— 2007b. 'Water advisory issued as B.C. flood watch expands'. 10 June. www.cbc.ca/canada/story/2007/06/10/bc-flooding.html.

——— 2008. 'Whiteout contributes to multi-car pileup on Ontario highway'. 20 January. www.cbc.ca/canada/story/2008/01/20/highway-pileup.html.

CBC Archives. 1976. '1912 Regina cyclone'. http://archives.cbc.ca/IDC-1-70-1713-11750/disasters_tragedies/tornadoes.

——— 1985. Broadcast, 3 June. Barrie tornado. http://archives.cbc.ca/IDC-1-70-1713-11757/disasters_tragedies/tornadoes/clip6.

CBC Radio One. 2007. Interview with Jacqueline Meszaros of the National Science Foundation, 'The Psychology of Risk'. 24 October.

Chagnon, S.A., R.A. Pielke Jr, D. Chagnon, R.T. Sylves, and R. Pulwarty. 2000. 'Human factors explain the increased losses from weather and climate extremes'. *Bulletin of the American Meteorological Society* 81 (3): 437–42.

Charleton, R., B. Kachman, and L. Wojtiw. 1995. 'Urban hailstorms: A view from Alberta'. *Natural hazards* 12:29–75.

CHC (Canadian Hurricane Centre). 2003. 'Impacts of hurricanes'. www.atl.ec.gc.ca/weather/hurricane/hurricanes3.html.

——— 2004. 'Canadian Hurricane Centre forecasting'. www.atl.ec.gc.ca/weather/hurricane/chc3.html#satellite.

——— 2006. 'What's in a name?' www.atl.ec.gc.ca/weather/hurricane/hurricanes4.html.

Chronicle-Herald. 1963. 'Nova Scotia escapes brunt of storm as Ginny lashes Nova Scotia'. 30 October.

——— 2003. 28 September.

——— 2006. 'Snow falls, traffic stalls'. 5 December, A1.

——— 2007a. 'Hurricane Juan: Then and now'. 29 September.

——— 2007b. 'Manitoba tornadoes renew calls for weather warning system'. 26 June.

——— 2007c. 'Not so sweet Easter surprise'. 9 April, A2.

——— 2007d. 'The power of Noel'. 5 November.

——— 2007d. 'The power of Noel'. 6 November.

——— 2008. 'Flood sweat and tears'. 2 May, A3.

Clean Air Strategic Alliance. 2007. 'Air Quality Index'. www.casadata.org/airqualityindex/index.asp.

Cohen, S., B. Bass, D. Etkin, B. Jones, J. Lacroix, B. Mills, D. Scott, and G.C. van Kooten. 2004. 'Regional adaptation strategies'. In *Hard choices: Climate change in Canada,* ed. H. Coward and A. Weaver, 151–78. Waterloo, ON: Wilfrid Laurier University Press.

Conrad, C., and R. Hanson. 2004. 'Actual versus perceived risk of hurricanes in Nova Scotia in the year prior to Hurricane Juan and the potential for perception change in a post-Juan period'. Paper presented at the Annual Meeting of the Canadian Association of Geographers, Moncton, NB.

CRIACC (Centre de ressources en impacts et adaption au climat et à ses changements). 2001. 'The snowstorm of the century'. Commemorative report of the snowstorm of 3–5 March 1971. Part of a series on extreme weather in Quebec.

Crichton, D. 1999. 'The Risk Triangle'. In *Natural Disaster Management,* ed. J. Ingleton, 102–3. London: Tudor Rose.

CTV (Canadian Television Network). 2006. 'Smog makes an early appearance in Canada'. http://toronto.ctv.ca/servlet/an/local/CTVNews/20060530/smog_laurie_060530?hub=TorontoHome.

——— 2007. 'Manitoba tornado was strongest ever'. www.ctv.ca/servlet/ArticleNews/story/CTVNews/20070918/manitoba_tornado_070918/20070918?hub=SciTech.

Cubasch, U., G.A. Meehl, G.J. Boer, R.J. Stouffer, M. Dix., A. Noda, C.A. Senior, S. Raper, and K.S. Yap. 2001. 'Projections of future climate change'. In *Climate change 2001: The scientific basis*. Contribution of Working Group I to the Third Assessment Report of the IPCC. Cambridge, UK, and New York: Cambridge University Press.

Danard, M., A. Munro, and T. Murty. 2003. 'Storm surge hazard in Canada'. *Natural Hazards* 28:407–31.

Davis, R.E., P.C. Knappenberger, W.M. Novicoff, and P.J. Michaels. 2002. 'Decadal changes in heat-related human mortality in the eastern United States'. *Climate Research* 22:175–84.

de Blij, H., P. Muller, R. Williams, C. Conrad, and P. Long. 2005. *Physical geography: The global environment*. Don Mills, ON: Oxford University Press.

Dey, B. 1982. 'Nature and possible causes of droughts on the Canadian Prairies—Case Studies'. *International Journal of Climatology* 2:233–49.

Dolney, T., and S. Sheridan. 2006. 'The relationship between extreme heat and ambulance response calls for the city of Toronto, Ontario, Canada'. *Environmental Research* 101:94–103.

Dore, M.H. 2003. 'Forecasting the conditional probabilities of natural disasters in Canada as a guide for disaster preparedness'. *Natural Hazards* 28 (2–3): 249–69.

Downer, Major General Hunt. 2007. 'The nation's hurricane program: An interagency success story'. A presentation at the US 61st Interdepartmental Hurricane Conference, 5–9 March 2007, New Orleans, LA.

Duneier, M. 2004. 'Scrutinizing the heat: On ethnic myths and the importance of shoe leather'. *Contemporary Sociology* 33 (2): 139–50.

Dunn, G.E., and B.I. Miller. 1960. *Atlantic hurricanes*. Baton Rouge, LA: Louisiana State University Press.

Easterling, D.R., H.F. Diaz, A.V. Douglas, W.D. Hogg, K.E. Kunkel, J.C. Rogers, and J.F. Wilkinson. 1999. 'Long-term observations for monitoring extremes in the Americas'. *Climatic Change* 42:285–308.

Easterling, D.R., J.L. Evans, P. Groisman, T.R. Karl, K.E. Kunkel, and P. Ambenje. 2000. 'Observed variability and trends in extreme climate events: A brief review'. *Bulletin of the American Meteorological Society* 81 (3): 417–25.

Emanuel, K. 2005. 'Increasing destructiveness of tropical cyclones over the past 30 years'. *Nature* 436:686–8.

Emergency Preparedness Canada. 2002. Office of Critical Infrastructure Protection and Emergency Preparedness. www.psepc-sppcc.gc.ca.

Environment Canada. 1987. *Climate atlas of Canada*. Map Series 3, Pressure, humidity, cloud, visibility, and days with thunderstorms, hail, smoke and haze, fog, freezing precipitation, blowing snow, frost, snow on the ground. Ministry of Supply and Services, Cat. No. EN56-63/3-1986.

———. 1996. 'Climate and weather glossary of terms'. Atmospheric, Climate, and Water Systems Branch. www.cmc.ec.gc.ca/climate/glossary.htm.

———. 1999. 'Top ten weather stories of 1999'. www.ec.gc.ca/press/10stor99_b_e.htm.

———. 2001. 'Blizzards—I am Canadian!' *Envirozine* 4 (1). www.ec.gc.ca/envirozine/english/issues/04/feature1_e.cfm.

———. 2002a. 'Severe weather watcher handbook'. www.msc-smc.ec.gc.ca/education/severe_weather/page02_e.htm.

———. 2002b. 'The worst ice storm in Canadian history?' www.msc-smc.ec.gc.ca/media/icestorm98/icestorm98_the_worst_e.cfm.

———. 2003a. 'Environment Canada's Wind Chill Program'. www.msc-smc.ec.gc.ca/education/windchill/index_e.cfm.

———. 2003b. 'Glossary of hurricane terms'. www.atl.ec.gc.ca/weather/hurricane/hurricanes9.html#tstorm.

——— 2003c. 'Lightning hot spots in Canada'. www.msc.ec.gc.ca/education/lightning/provinces_e.html.

——— 2003d. 'Natural disasters on the rise'. *Science and the Environment Bulletin*, March/April.

——— 2003e. 'Top Canadian weather winners'. www.on.ec.gc.ca/weather/winners/winners-e.html.

——— 2004a. 'Causes of flooding—Coastal storms'. www.ec.gc.ca/WATER/en/manage/floodgen/e_storm.htm.

——— 2004b. 'Causes of flooding—Outburst floods'. www.ec.gc.ca/WATER/en/manage/floodgen/e_outbur.htm.

——— 2004c. 'Causes of flooding—Snowmelt runoff'. www.ec.gc.ca/WATER/en/manage/floodgen/e_snow.htm.

——— 2004e. 'Causes of flooding—Urban stormwater runoff'. www.ec.gc.ca/WATER/en/manage/floodgen/e_urban.htm.

——— 2004f. 'The management of water'. www.ec.gc.ca/water/en/manage/floodgen/e_icejam.htm.

——— 2004g. 'Remembering Hurricane Hazel'. www.atl.ec.gc.ca/weather/hurricane/hazel/en/index.html.

——— 2004h. 'Severe weather forecasting improves with completion of Canada's National Doppler Radar Project'. News release, 29 September. www.ec.gc.ca/press/2004/040929_n_e.htm.

——— 2004i. 'Top weather events of the 20th century'. www.ec.gc.ca/press/vote2-_f_e.htm.

——— 2004j. 'White Juan'. www.atl.ec.gc.ca/weather/severe/2003-2004/whitejuan_e.html.

——— 2005a. 'Atmospheric hazards web site: Ontario'. http://ontario.hazards.ca/historical/Fog_Ontario-e.html.

——— 2005b. CANWARN fact sheet. www.on.ec.gc.ca/severe-weather/canwarn-e.html.

——— 2005c. 'A climatology of hurricanes for Canada: Improving our awareness of the threat'. Meteorological Service of Canada.

——— 2005d. 'Severe summer weather'. www.mb.ec.gc.ca/air/summersevere/ae00s16.en.html.

——— 2005e. 'Severe summer weather in Alberta'. www.mb.ec.gc.ca/air/summersevere/ae00s16.en.html.

——— 2005f. 'Severe winter weather events'. www.mb.ec.gc.ca/air/wintersevere/events.en.html.

——— 2006a. 'Canadian climate normals or averages 1971–2000'. www.climate.weatheroffice.ec.gc.ca/climate_normals/index_e.html.

——— 2006b. 'The Climate of Newfoundland'. http://atlantic-web1.ns.ec.gc.ca/climatecentre/default.asp?lang=En&n=83846147- 1.

——— 2006c. 'Humidity'. www.msc-smc.ec.gc.ca/cd/brochures/humidity_e.cfm.

——— 2006d. 'Severe weather: Issuing weather warnings to protect the safety and security of Canadians'. *Science for Sustainable Development 7*. Government of Canada fact sheet. http://badc.nerc.ac.uk/community/poster_heaven_old/Poster07E.pdf.

——— 2006e. 'Summer 2006'. www.msc-smc.ec.gc.ca/ccrm/bulletin/national_e.cfm.

——— 2006f. 'Top weather events of the 20th century'. www.ec.gc.ca/press/vote20_f_e.htm.

——— 2006g. 'Weather watches, warnings and advisories'. www.msc-smc.ec.gc.ca/cd/brochures/warning_e.cfm.

——— 2007a. 'Air issues in Atlantic Canada'. www.ec.gc.ca/cleanair-airpur/Atlantic_Region-WSE8DA13D2-1_En.htm.

——— 2007b. Canadian Hurricane Centre. www.atl.ec.gc.ca/weather/hurricane/connection.html.

——— 2007c. 'The climate of Nova Scotia'. http://atlantic-web1.ns.ec.gc.ca/climatecentre.

——— 2007d. 'Fact sheet: What is ground-level ozone?' http://lavoieverte.qc.ec.gc.ca/atmos/smog/fiche_info_e.html.

——— 2007e. 'Fraser Valley smog: An indicator of potential air quality health risk'. www.ecoinfo.org/env_ind/region/smog/smog_e.cfm.

——— 2007f. 'Hurricane Juan'. www.atl.ec.gc.ca/weather/hurricane/juan.

——— 2007g. 'Monthly and seasonal forecasts'. www.weatheroffice.gc.ca/saisons/index_e.html.

——— 2007h. 'Post-tropical storm Noel batters Eastern Canada'. *Envirozine*, no. 77 www.ec.gc.ca/EnviroZine/english/issues/77/nature_e.cfm.

——— 2007i. 'Reducing flood damage'. www.ec.gc.ca/water/en/manage/floodgen/e_red_fr.htm.

——— 2007j. 'Thunderstorms'. www.mb.ec.gc.ca/air/summersevere/ae00s20.en.html.

——— 2007k. 'Winter 2006/07'. www.msc-smc.ec.gc.ca/ccrm/bulletin/national_e.cfm.

——— 2008a. 'Canada's top ten weather stories for 2007'. www.environment-canada.ca/doc/smc-msc/m_110/toc_eng.html.

——— 2008b. 'Lightning' Weather Office, www.weatheroffice.gc.ca/lightning/index_e.html.

EPA (Environmental Protection Agency). 2008. 'Heat Island Effect'. www.epa.gov/hiri.

Etkin, D. 1995. 'Beyond the year 2000, more tornadoes in Western Canada? Implications from the historical record'. *Natural Hazards* 12:19–27.

——— 1997. 'The social and economic impact of hydrometeorological hazards and disasters: A preliminary inventory'. In *Coping with natural hazards in Canada: Scientific, government and insurance industry perspectives*, ed. S.E. Brun, D. Etkin, D. Gesink Law, L. Wallace, and R. White. A study written for the Round Table on Environmental Risk, Natural Hazards and the Insurance Industry.

——— 1999. 'Risk transference and related trends: Driving forces towards more mega-disasters'. *Environmental Hazards* 1:69–75.

Etkin, D., and S. Brun. 2001. 'Canada's hail climatology: 1977–1993'. Institute for Catastrophic Loss Reduction Paper Series No. 14.

Etkin, D., and I. Leman Stefanovic. 2005. 'Mitigating natural disasters: The role of eco-ethics'. In *Mitigation of natural hazards and disasters: International perspectives*, ed. C.E. Haque, 135–58. Dordrecht, The Netherlands: Springer.

Etkin, D., and A. Maarouf. 1995. 'An overview of atmospheric natural hazards in Canada'. In *Proceedings of a Tri-lateral Workshop on Natural Hazards*, ed. D. Etkin, 1-63–1-92. Merrickville, ON, 11–14 February.

Etkin, D, and M. Myers. 2000. 'Thunderstorms in a social context'. In *Storms,* ed. R. Pielke Jr, and R. Pielke, Sr, 2:43–59. New York, NY: Routledge.

Etkin, D., S. Brun, A. Shabbar, and P. Joe. 2001. 'Tornado climatology of Canada revisited: Tornado activity during different phases of ENSO'. *International Journal of Climatology* 21:915–38.

Etkin, D., S. Brun, S. Chrom, and P. Dogra. 2002. 'A tornado scenario for Barrie, Ontario'. ICLR Research Paper Series, No. 20.

Etkin, D., E. Haque, L. Bellisario, and I. Burton. 2004. 'An assessment of natural hazards and disasters in Canada: A report for decision-makers and practitioners'. Canadian Natural Hazards Assessment Project.

Extreme Science. 2008. www.extremescience.com/wettest.htm.

Fitzharris, B.B. 1987. ' A climatology of major avalanche winters in western Canada'. *Atmosphere-Ocean* 25 (2): 115–36.

Foreign Affairs and International Trade Canada. 2006. *Highlights of Canadian disaster reduction activities*. Government of Canada. www.dfait-maeci.gc.ca.

Francis, D., and H. Hengeveld. 1998. *Extreme weather and climate change*. Climate and Water Products Division, Atmospheric Environment Service, ON.

Galbraith, P.W. 1980. 'Hurricanes and tropical storms in Atlantic Canada—A brief review'. Internal Report #MAES 6-80, August. Environment Service, Bedford, NS.

Geis, D. 2000. 'By design: The disaster-resistant and quality-of-life community'. *Natural Hazards Review* 1 (3): 151–60.

Gerber, B., and G. Neeley. 2005. 'Perceived risk and citizen preferences for governmental management of routine hazards'. *Policy Studies Journal* 33 (3): 395–418.

Global Change Strategies International Inc. 2000. *Water sector: Vulnerability and adaptation to climate change*. Report. Meteorological Service of Canada.

Globe and Mail. 2006a. 'B.C. coasters can drink the water if it isn't frozen'. 27 November. www.theglobeandmail.com/servlet/story/RTGAM.20061127.wboilwater11 27.

——— 2006b. 'Wild B.C. weather leaves 200,000 without power'. Posted online on 16 November 2006. Accessed 28 November 2006 at www.theglobeandmail.com/servlet/Page/document/v5/content/subscribe?us er_URL=www.theglobeandmail.com%2Fservlet%2Fstory%2FLAC.200611 16.BCRAIN16%2FTPStory%2FNational&ord=4916967&brand=theglobeandmai l&force_login=true.

——— 2007. 'Southern B.C. awash in pineapple juice'. 11 June. www.theglobeandmail.com/servlet/story/RTGAM.20061106.wsoggy1106.

Goldenberg, S.B., C.W. Landsea, A.M. Mesatas-Nuñez, and W.M. Gray. 2001. 'The recent increase in Atlantic hurricane activity: Causes and implications'. *Science* 293:474–9.

Government of Australia. 2008. Department of Foreign Affairs and Trade: Travel advisories: Canada. www.smartraveller.gov.au/zw- cgi/view/Advice/Canada.

Government of Canada. 2002. 'Introduction: Smog and health'. Parliamentary Research Branch.

Gray, W.M., J.D. Sheaffer, and C.W. Landsea. 1997. 'Climate trends associated with multidecadal variability of Atlantic hurricane activity'. In *Hurricanes: Climate and socioeconomic impacts,* ed. H.F. Diazand and R.S. Pulwarty, 15–53. New York: Springer-Verlag.

Green Ontario. 2007. 'Smog'. www.greenontario.org/strategy/smog.html.

Greene, W., ed. 1998. *Proceedings of the Mitigation Symposium: Towards a Canadian national mitigation strategy*. Vancouver: University of British Columbia Disaster Preparedness Resources Centre.

Groisman, P., and D Easterling. 1993. 'Variability and trends of total precipitation and snowfall over the United States and Canada'. *Journal of Climate* 7:184–205.

Groisman, P.Y., R.W. Knight, T.R. Karl, D.R. Easterling, B.M. Sun, and J.H. Lawrimore. 2004. 'Contemporary changes of the hydrological cycle over the contiguous United States: Trends derived from in situ observations. *Journal of Hydrometeorology* 5:64–85.

Gulev, S., O. Zolina, and S. Grigoriev. 2001. 'Winter storms in the Northern Hemisphere (1958–1999)'. *CO2 Science*. www.co2science.org/articles/V4/N37/C2.php.

Gultepe, I., R. Tardif, S.C. Michaelides, J. Cermak, A. Bott, J. Bendix, M.D. Müller, M. Pagowski, B. Hansen, G. Ellrod, W. Jacobs, G. Toth, and S.G. Cober. 2007. 'Fog research: A review of past achievements and future perspectives'. *Pure and Applied Geophysics* 164:1121–59.

Guttman, N. 1998. 'Comparing the Palmer Drought Index and the Standard Precipitation Index'. *Journal of American Water Resources Association* 34 (1): 113–21.

Hanesiak, J., and X. Wang. 2005. 'Adverse-weather trends in the Canadian Arctic'. *Journal of Climate* 18 (16): 3140–56.

Hanson, B. 2006. 'Study of fog climatology of Canada'. http://collaboration.cmc.ec.gc.ca/science/arma/climatology/regions.html#_Refs#_ Refs.

——— 2002. 'Actual versus perceived risk of hurricanes in Nova Scotia'. Undergraduate honours thesis, Department of Geography, Saint Mary's University, Halifax, NS.

——— 2005. 'Hurricane Juan: A survey of impacts and vulnerability'. Master's thesis, Department of Geography, University of Toronto.

Haque, C.E. 2005. *Mitigation of natural hazards and disasters*. Dordrecht, The Netherlands: Springer.

Haque, C.E., and Kilgour, J. 2000. 'The Canadian Natural Hazards Assessment: A report on the First Workshop'. Emergency Preparedness Canada. Workshop held at University of Manitoba.

Hayhoe, K., D. Cayan, C.B. Field, P.C. Frumhoff, E.P. Maurer, N.L. Miller, S.C. Moser, S.H. Schneider, K. Nicholas Cahill, E.E. Cleland, L. Dale, R. Drapek, R.M. Hanemann, L.S. Kalkstein, J. Lenihan, C.K. Lunch, R.P. Neilson, S.C. Sheridan, and J.H. Verville. 2004. 'Emission pathways, climate change, and impacts on California'. *Proceedings of the National Academy of Sciences of the United States of America* 101 (34): 12422–7.

Hearn Morrow, B. 1999. 'Identifying and mapping community vulnerability'. *Disasters* 23 (10): 1–18.

Heidorn, K. 2000. 'The days Niagara Falls went dry'. *The Weather Doctor's Weather Almanac*. www.islandnet.com/~see/weather/almanac/arc2000/alm00mar.htm.

——— 2002. 'The fog rolls in'. www.islandnet.com/~see/weather/almanac/arc2002/alm02sep.htm.

——— 2004. 'Wreckhouse winds'. www.islandnet.com/~see/weather/almanac/arc2005/alm05dec.htm.

——— 2005a. *'And now . . . the weather'*. Calgary: Fifth House.

——— 2005b. 'Storm surge'. www.islandnet.com/~see/weather/almanac/arc2004/alm04sep.htm.

Henderson-Sellers, A., H. Zhang, G. Berz, K. Emanuel, W. Gray, C. Landsea, G. Holland, J. Lighthill, S.L. Shieh, P.Webster, and K. McGuffie. 1998. 'Tropical cyclones and global climate change: A post-IPCC assessment'. http://citeseer.ist.psu.edu/470194.html.

Henson, B. 2005. 'Going to extremes'. *UCAR Quarterly*, Winter 2004–5. Also available online at www.ucar.edu/communications/quarterly/winter04/extremes.html.

Henstra, D. 2003. 'Federal emergency management in Canada and the United States after 11 September 2001'. *Canadian Public Administration* 46 (1): 103–16.

Henstra, D., and G. McBean, 2004. *The role of government in services for disaster mitigation*. London, ON: Institute for Catastrophic Loss Reduction.

Hogg, W.D., and D.A. Carr. 1985. *Rainfall frequency atlas for Canada*. Ottawa: Supply and Services Canada.

Houghton, J.T., Y. Ding, D.J. Griggs, M. Noguer, P.J. van der Linden, D. Xiaosu, K. Maskell, and C. Johnson, eds. 2001. *Climate change 2001: The scientific basis*. Contribution of Working Group I to the Third Assessment Report of the Intergovernmental Panel on Climate Change. Cambridge, UK: Cambridge University Press.

Houtekamer, P.L., L. Lefaivre, J. Derome, H. Ritchie, and H.L. Mitchell. 1996. 'A system simulation approach to ensemble prediction'. *Monthly Weather Review* 124:1225–42.

Hwacha, V. 2005. 'Canada's experience in developing a national disaster mitigation strategy: A deliberative dialogue approach'. *Mitigation and Adaptation Strategies for Global Change* 10 (3): 507–23.

ICLR (Institute for Catastrophic Loss Reduction). 1998. *Better protecting Canadians from natural hazards*. Toronto: ICLR.

Insurance Bureau of Canada. 2006. *Natural disaster reduction plan*. www.ibc.ca/ii_natural.asp.

IPCC (Intergovernmental Panel on Climate Change). 2007a. *Climate change 2007: Impacts, adaptation and vulnerability*, ed. M.L. Parry, O.F. Canziani, J.P. Palutikof, P.J. van der Linden, and C.E. Hanson. Contribution of Working Group II to the Fourth Assessment Report of the IPCC. Cambridge, UK: Cambridge University Press.

——— 2007b. 'Climate change 2007: The Physical Basis of Climate Change'. Contribution of Working Group I to the Fourth Assessment Report of the IPCC.

IQP. 2007. 'At the cutting edge: pale green'. *Inside Queen's Park* 20 (10).

Jaeger, C., O. Renn, E. Rosa, and T. Webler. 2002. *Risk and rational action*. London: Earthscan.

Jones, B., and J. Andrey. 1998. 'Weather warnings and adaptive responses: Perceptions of Kingston, Ontario residents'. University of Waterloo paper presented at the Adaptation Learning Project Workshop, 12–13 November, Downsview, ON.

Jones, R.L. 1992. 'Canadian disasters—An historical survey'. *Natural Hazards* 5 (1): 43–51.

Kalkstein, L.S., and J.A. Greene. 1997. 'An evaluation of climate/mortality relationships in large U.S. cities and the possible impacts of a climate change'. *Environmental Health Perspectives* 105:84–93.

Karl, T.R., and D.R. Easterling. 1999. 'Climate extremes: Selected review and future research directions'. *Climatic Change* 42:309–25.

Khandekar, M. 2002. *Trends and changes in extreme weather events: An assessment with focus on Alberta and Canadian Prairies.* Environment Canada, Pub. No. I/927.

King, D. 2000. 'You're on your own: Community vulnerability and the need for awareness and education for predictable natural disasters'. *Journal of Contingencies and Crisis Management* 8 (4): 223–7.

Klinenberg, Eric. 2002. *Heat wave: A social autopsy of disaster in Chicago.* Chicago, IL: University of Chicago Press.

Klotzbach, P., and W.M. Gray. 2007. 'Extended range forecast of the Atlantic season hurricane activity and U.S. landfall strike probability for 2007'. http://hurricane.atmos.colostate.edu/Forecasts/2007/april2007.

Knowles, N., M.D. Dettinger, and D.R. Caya. 2006. 'Trends in snowfall versus rainfall in the western United States'. *Journal of Climate* 19:4545–59.

Knutson, T.R., and R.E. Tuleya. 2004. 'Impact of CO_2-induced warming on simulated hurricane intensity and precipitation: Sensitivity to the choice of climate model and convective parameterization'. *Journal of Climate* 17:3477–95.

Kocin, P.J. 1983. 'An analysis of "The Blizzard of '88"'. *Bulletin of the American Meteorological Society* 64 (11): 1258–72.

Kossin, J.P., K. Knapp, D. Vimont, R. Murname, and B. Harper. 2007. 'A globally consistent reanalysis of hurricane variability and trends'. *Geophysical Research Letters* 34, L04815.

Kossin, J.P., and D.J. Vimont. 2007. 'A more general framework for understanding Atlantic hurricane variability and trends'. *Bulletin of the American Meteorological Society* 88 (11): 1767–80.

Kovats, R.S., and K.L. Ebi. 2006. 'Heatwaves and public health in Europe'. *European Journal of Public Health* 16 (6): 592–9.

Kunkel, K. 2003. 'North American trends in extreme precipitation'. *Natural Hazards* 29:291–305.

Kunkel, K.E., K. Andsager, and D.R. Easterling. 1999a. 'Long-term trends in extreme precipitation events over the conterminous United States and Canada'. *Journal of Climate* 12:2515–27.

Kunkel, R.A., R.A. Pielke Jr, and S.A. Changnon. 1999b. 'Temporal fluctuations in weather and climate extremes that cause economic and human health impacts: A review'. *Bulletin of the American Meteorological Society* 80:1077–98.

Lambert, S.J. 1995. 'The effect of enhanced greenhouse warming on winter cyclone frequencies and strengths'. *Journal of Climate* 8:1447–52.

——— 2004. 'Changes in winter cyclone frequencies and strengths in transient enhanced greenhouse warming simulations using two coupled climate models'. *Atmosphere-Ocean* 42 (3): 173–81.

Landsberg, H.E. 1981. *The urban climate.* New York: Academic Press.

Landsea, C.W., G.D. Bell, W.M. Gray, and S.B. Goldenberg. 1998. 'The extremely active 1995 Atlantic hurricane season: Environmental conditions and verification of seasonal forecasts'. *Monthly Weather Review* 126:1174–93.

Landsea, C.W., R.A. Pielkel Jr, A.M. Mestas-Nuñez, and J.A. Knaff. 1999. 'Atlantic basin hurricanes: Indices of climate changes'. *Climatic Change* 42:89–129.

Landsea, C., J. Franklin, C. McAdie, J. Beven II, J. Gross, B. Jarvinen, R. Pasch, E. Rappaport, J. Dunion, and P. Dodge. 2004. 'A reanalysis of Hurricane Andrew's intensity'. *Bulletin of the American Meteorological Society* 85 (11): 1699–712.

Lawford, R., T. Prowse, W. Hogg, A. Warkentin, and P. Pilon. 1995. 'Hydrometeorological aspects of flood hazards in Canada', *Atmosphere-Ocean* 33 (2): 303–20.

Lawrimore, J., R. Heim, M. Svoboda, V. Swail, and P. Englehart. 2002. 'Beginning a new era of drought monitoring across North America'. *Bulletin of the American Meteorological Society* 83 (8): 1191–2.

Lawson, B. 1987. 'The climatology of blizzards in Western Canada 1953–1986'. Central Region Report CAES 88-1. Atmosphere Environment Service, Winnipeg.

Lawson, B. 2003a. 'Trends in winter extreme minimum temperatures on the Canadian Prairies'. *Atmosphere-Ocean* 41 (3): 233–9.

Lawson, B. 2003b. 'Trends in blizzards at selected locations on the Canadian Prairies'. *Natural Hazards* 29:123–38.

Lecomte, E., A. Pang, and J. Russell. 1998. *Ice storm '98*. ICLR Research Paper Series, No. 1.

Leiss, W., and C. Chociolko. 1994. *Risk and responsibility*. Montreal and Kingston: McGill-Queen's University Press.

Lines, G. 2007. Head, Climate Change Section, Meteorological Service of Canada. Personal communication.

Lott, N. 1993. 'The Big One! A review of the March 12–14, 1993 'Storm of the Century'. National Climatic Data Center Technical Report 93-01.

McBean, G.A. 2005. 'Risk mitigation strategies for tornadoes in the context of climate change and development'. *Mitigation and adaptation strategies for global change* 10:357–366.

McBean, G., and D. Henstra. 2003. 'Climate change, natural hazards and cities'. Natural Resources Canada.

McEntire, D.A. 2001. 'Triggering agents, vulnerabilities and disaster reduction'. *Disaster Prevention and Management* 10 (3): 189–96.

——— 2005. 'Why vulnerability matters: Exploring the merit of an inclusive disaster reduction concept'. *Disaster Prevention and Management* 14 (2): 206–22.

McInnes, K.L., K. Walsh, G. Hubbert, and T. Beet. 2003. 'Impact of sea-level rise and storm surges on a coastal community'. *Natural Hazards* 30:187–207.

McIntyre, C. 2006. 'Defogging the mysteries of extreme weather'. In *McGill Tribune*, online edition, 4 April. www.mcgilltribune.com/media/storage/paper234/news/2002/03/18/Features/Defogging.The.Mysteries.Of.Extreme.Weather-218846.shtml?norewrite 200605192057&sourcedomain=www.mcgilltribune.com.

McLeod, D. 2003. *Hurricane Juan: The unforgettable storm*. Halifax: Formac Publishing Company.

Manitoba Conservation Air Quality Section. 2007. 'What is smog?' www.gov.mb.ca/conservation/airquality/smog.html.

Manitoba Floodway Authority. 2008. 'What is floodway expansion?' www.floodwayauthority.mb.ca.

Masters, Jeff. 2006. 'Flying into a record nor'easter'. Wunderground.com. www.wunderground.com/blog/JeffMasters/comment.html?entrynum=304& tstamp=200602.

Maybank, J., B.R. Bonsal, K. Jones, R.G. Lawford, E.G. O'Brien, A. Ripley, and E. Wheaton. 1995. 'Drought as a natural disaster'. *Atmosphere-Ocean* 33:195–222.

Meehl, G.A., and C. Tebaldi. 2004. 'More intense, more frequent, and longer lasting heat waves in the 21st century'. *Science* 305:994–7.

Meehl, G.A., T. Karl, D.R. Easterling, S. Chagnon, R. Pielke Jr, D. Changnon, J. Evans, P. Groisman, T.R. Knutson, K.E. Kunkel, L.O. Mearns, C. Parmesan, R. Pulwarty, T. Root, R.T. Sylves, P. Whetton, and F. Zwiers. 2000a. 'An introduction to trends in extreme weather and climate events: Observations, socioeconomic impacts, terrestrial ecological impacts, and model projections'. *Bulletin of the American Meteorological Society* 81 (3): 413–16.

Meehl, G.A., F. Zwiers, J. Evans, T. Knutson, L. Mearns, and P. Whetton. 2000b. 'Trends in extreme weather and climate events: Issues related to modeling extremes in projections of future climate change'. *Bulletin of the American Meteorological Society* 81 (3): 427–36.

Mekis, E., and W. Hogg. 1999. 'Rehabilitation and analysis of Canadian daily precipitation time series'. *Atmosphere-Ocean* 37 (1): 53–85.

Mickley, L.J. 2007. 'A future short of breath? Possible effects of climate change on smog'. *Environment* 49 (6): 32–43.

Mills, B., C. Parkinson, B. Jones, J. Yessis, and K. Spring. 2008. 'Assessment of lightning-related fatality and injury risk in Canada'. *Natural Hazards.* SpringerLink. www.springerlink.com/content/1107360702228010.

Mills, E. 2005. 'Insurance in a climate of change'. *Science* 309:1040–4.

Ministry of Ontario. 'Road salt management'. www.mto.gov.on.ca/english/engineering/roadsalt.htm.

Moran, J., and M. Morgan. 1997. *Meteorology: The atmosphere and the science of weather.* 5th ed. Upper Saddle River, NJ: Prentice Hall.

Moran, J., M. Morgan, and P. Pauley. 1997. *Meteorology: The atmosphere and the science of weather.* Upper Saddle River, NJ: Prentice Hall.

MSC (Meteorological Service of Canada). 2002. 'The evolution of Doppler radar'. www.msc-smc.ec.gc.ca/projects/nrp/evolution_e.cfm.

——— 2005. 'The Canadian Lightning Detection Network: Novel approaches for performance measurement and network management', by D. Dockendorff and K. Spring. Paper presented at the World Meteorological Organization (WMO) Technical Conference on Meteorological and Environmental Instruments and Methods of Observation (TECO 2005), Bucharest, Romania, 4–7 May 2005.

Muraca, G., D.C. MacIver, N. Urquizo, and H. Auld. 2001. 'The climatology of fog in Canada'. In *Proceedings of the Second International Conference on Fog and Fog Collection,* 513–16. St John's, NL.

NASA. 2005. 'In search of the driest place on Earth'. http://quest.nasa.gov/challenges/marsanalog/egypt/index.html.

National Hurricane Center. 2008. 'Tropical cyclone climatology'. www.nhc.noaa.gov/pastprofile.shtml.

National Radar Program. 2005. 'Implementation of the National Radar Project'. Environment Canada. www.msc-smc.ec.gc.ca/projects/nrp/answers2_e.cfm#2-1.

National Science Foundation. 2007. 'Risking Wildfire'. www.nsf.gov.

National Severe Storms Laboratory. 2006. 'Forecast and warning improvements'. www.nssl.noaa.gov/research/forewarn.

Natural Resources Canada. 2004. 'Discover Canada through maps and facts'. http://atlas.nrcan.gc.ca/site/english/learningresources/quizzes/download.html?category=3.

——— 2006. 'Flood disasters in Canada'. http://gsc.nrcan.gc.ca/floods/database_e.php.

——— 2007. 'From impacts to adaptation: Canada in a changing climate 2007'. http://adaptation.rncan.gc.ca/assess/2007/synth/intro_e.php.

Nav Canada. 2008. 'Weather patterns of British Columbia'. www.navcanada.ca/NavCanada.asp?Language=en&Content=ContentDefinitionFiles\Publications\LAK\default.xml.

NCDC (National Climate Data Center). 2005. 'Climate of 2005: Annual report'. www.ncdc.noaa.gov/oa/climate/research/2005/ann/global.html.

Neumann, C.J., G.W. Cry, E.L. Caso, and B.R. Jarvinen. 1978. *Tropical cyclones of the North Atlantic Ocean, 1871–1977.* Ashveille, NC: NOAA.

Newark, M. 1984. 'Canadian tornadoes, 1950–1979'. *Atmosphere-Ocean* 22 (3): 343–53.

Newton, J. 1997. 'Federal legislation for disaster mitigation: A comparative assessment between Canada and the United States'. *Natural Hazards* 16:219–41.

────── 2002. *Towards a national mitigation policy: An investigation of efforts to create safer communities—Experiences in Canada and the United States*. Ottawa: Office of Critical Infrastructure Protection and Emergency Preparedness, Government of Canada.

Newton, J., J. Paci, and A. Ogden. 2005. 'Climate change and natural hazards in northern Canada: Integrating indigenous perspectives with government policy'. In *Mitigation of natural hazards and disasters: International perspectives*, ed. C.E. Haque, 209–39. Dordrecht, The Netherlands: Springer.

Nkemdirim, L., and L. Weber. 1999. 'Comparison between the droughts of the 1930s and the 1980s in the southern Prairies of Canada'. *Journal of Climate* 12:2434–50.

NOAA (National Oceanic and Atmospheric Organization). 1995. *Surface weather observations and reports*. Federal Meteorological Handbook, No. 1.

────── 1999. 'Hurricane basics'. http://hurricanes.noaa.gov/pdf/hurricanebook.pdf.

────── 2003. *Hurricane Modification*. Atlantic Oceanographic and Meteorological Laboratory. www.aoml.noaa.gov/hrd/hrd_sub/modification.html.

────── 2006. 'Climate of 2005: Atlantic hurricane season'. www.ncdc.noaa.gov/oa/climate/research/2005/hurricanes05.html#sum.

────── 2007a. 'Atlantic Hurricane Season Outlook'. www.cpc.noaa.gov/products/outlooks/hurricane.shtml.

────── 2007b. 'History of tornado forecasting'. http://celebrating200years.noaa.gov/magazine/tornado_forecasting/welcome.html#today/.

────── 2007c. 'North American Drought Monitor'. www.ncdc.noaa.gov/oa/climate/monitoring/drought/nadm.

────── 2007d. 'The Online Tornado FAQ'. www.spc.noaa.gov/faq/tornado/index.html.

────── 2007e. 'The Saffir-Simpson Hurricane Scale'. www.nhc.noaa.gov/aboutsshs.html.

────── 2007f. 'Volcanic lightning'. National Geophysical Data Center.

────── 2007g. 'Worldwide tropical cyclone names'. www.nhc.noaa.gov/aboutnames.shtml.

────── 2008. 'Frequently asked questions'. www.aoml.noaa.gov/hrd/tcfaq/A6.html.

North Atlantic Oscillation. www.ldeo.columbia.edu/res/pi/NAO.

Nott, M.P., D.F. Desante, R.B. Siegel, and P. Pyle. 2002. 'Influences of the El Niño/Southern Oscillation and the North Atlantic Oscillation on avian productivity in forests of the Pacific Northwest of North America'. *Global Ecology and Biogeography Letters* 11 (4): 333–42.

Oke, T.R. *Boundary layer climates*. New York: Routledge.

Oke, T.R., and R.F. Fuggle. 1972. 'Comparison of urban/rural counter and net radiation at night'. *Boundary-Later Meteorology* 2:290–308.

Ontario Ministry of Agriculture, Food and Rural Affairs. 2007. 'Extreme cold temperatures over the past winter likely to cause winter injury to apple trees'. www.omafra.gov.on.ca/english/crops/hort/news/orchnews/2005/on_0605a6.htm.

Pagowski, M., I. Gultepe, and P. King. 2004. 'Analysis and modeling of an extremely dense fog event in southern Ontario'. *Journal of Applied Meteorology* 43:3–16.

Paul, A.H. 1995a. 'The Saskatchewan Tornado Project'. University of Regina, Department of Geography Internal Report.

PBS (Public Broadcasting Service). 2007. www.pbs.org/lostliners/empress.html.

Peel, S., and L. Wilson. 2006. 'Precipitation forecasts of the Canadian Ensemble Prediction System'. In *Proceedings of the 18th Conference on Probability and Statistics in the Atmospheric Sciences*. Atlanta, GA: American Meteorological Society.

Phillips, D. 1990. *The climates of Canada*. Environment Canada, Supply and Services Canada Publishing Centre, Cat. No. EN56-1/1990E.

────── 1997. 'The top 10 weather stories of 1996'. *Bulletin, Canadian Meteorological and Oceanographic Society* 25:1–2.

Phillips, D.W., and R.B. Crowe. 1984. *Climate Severity Index for Canadians*. Environment Canada, Atmospheric Environment Service.

Pielke, R.A., R. Klein, and D. Sarawitz. 2000. 'Turning the big knob: Energy policy as a means to reduce weather impacts'. *Energy and the Environment* 11:255–76.

Pielke, R.A., Jr, C. Landsea, M. Mayfield, J. Laver, and R. Pasch. 2005. 'Hurricanes and global warming'. *Bulletin of the American Meteorological Society* 86 (11): 1571–5.

Province (Vancouver). 2008. 'Black ice causes chain-reaction crash involving fire trucks'. 7 February. www.canada.com/theprovince/news/story.html?id=1c14fb23-2f22-400a-b657-82a7040b05fd&k=36945.

PSC (Public Safety Canada). 2005a. 'Canadian disaster database'. http://securitepublique.gc.ca/res/em/cdd/search-en.asp.

——— 2005b. 'Tornadoes'. www.ps-sp.gc.ca/res/em/nh/to/index-en.asp.

——— 2006a. 'Canadian disaster database'. www.publicsafety.gc.ca.

——— 2006b. 'Towards a National Disaster Mitigation Strategy'. http://ww3.psepc-sppcc.gc.ca/NDMS/index_e.asp.

——— 2007a. 'Hail'. www.ps-sp.gc.ca/res/em/nh/ha/index-en.asp.

——— 2007b. 'Icebergs, sea ice and fog'. www.ps-sp.gc.ca/res/em/nh/isif/index-eng.aspx.

——— 2007c. 'Significant floods of the 19th and 20th centuries'. www.publicsafety.gc.ca/res/em/nh/fl/fl-sig-en.asp.

Raddatz. R.L., and J.M. Hanesiak. 2008. 'Significant summer rainfall in the Canadian Prairie provinces: Modes and mechanisms 2000–2004'. *International Journal of Climatology*. Online ed. http://www3.interscience.wiley.com/journal/117905230/abstract?CRETRY=1&SRETRY=0.

Redner, S., and M. Petersen. 2006. 'On the role of global warming on the statistics of record-breaking temperatures'. *Atmospheric and Oceanic Physics*. http://arxiv.org/abs/physics/0509088v6.

Regina Plains Museum. 2006. 'A window into the Regina Tornado of 1912'. www.virtualmuseum.ca/pm.php?id=story_line&lg=English&fl=0&ex=0000 00306&sl=6697&pos=1.

Renn, O. 2004. 'Perceptions of risk'. *Toxicology Letters* 149 (1–3): 405–13.

Richardson, D. 2000. 'Skill and economic value of the ECMWF ensemble prediction system'. *Quarterly Journal of the Royal Meteorological Society* 126:649–68.

Ripley, E. 1986. 'Is the prairie becoming drier?' In *Drought: The impending crisis? Proceedings of the 16th Canadian Hydrology Symposium*, 50–60. Ottawa: National Research Council.

Robert, B., S. Forget, and J. Rouselle. 2003. 'The effectiveness of flood damage reduction measures in the Montreal region'. *Natural Hazards* 28:367–85.

Roberts, E., R. Stewart, and C. Lin. 2006. 'A study of drought characteristics over the Canadian prairies'. *Atmosphere-Ocean* 44 (4): 331–45.

Rouse, W.R., D. Noad, and J. McCutcheon. 1973. 'Radiation, temperature and atmospheric emissivities in a polluted urban atmosphere in Hamilton, Ontario'. *Journal of Applied Meteorology* 12:798–807.

Roy, E., J. Rousselle, and J. Lacroix. 2003. 'Flood damage reduction program (FDRP) in Quebec: Case study of the Chaudière River'. *Natural Hazards* 28:387–405.

Ruffman, A. 1995. Comment on 'The Great Newfoundland Storm of 12 September 1775', by Anne E. Stevens and Michael Staveley. *Bulletin of the Seismological Society of America* 85 (2): 646–9.

Sarawitz, D., R.A. Pielke Jr, and M. Keykhah. 2003. 'Vulnerability and risk: Some thoughts from a political and policy perspective'. *Risk Analysis* 23 (4): 805–10.

Saskatchewan Agriculture, Food and Rural Revitalization. 2007. 'Frost damage to crops'. www.agriculture.gov.sk.ca/Default.aspx?DN=0443309f-7978-42d5-ada9- 5d1fc0900fe5.

Shabbar, A., and W. Skinner. 2004. 'Summer drought patterns in Canada and the relationship to global sea surface temperatures'. *Journal of Climate* 17:2866–80.

Shabbar, A., B. Bonsal, and M. Khandekar. 1997. 'Canadian precipitation patterns associated with the Southern Oscillation'. *Journal of Climate* 10:3016–27.

Sheridan, S. 2006. 'A survey of public perception and response to heat warnings across four North American cities: An evaluation of municipal effectiveness'. *International Journal of Biometeorology* (online).

Sheridan, S., and L. Kalkstein. 1998. 'Heat watch-warning systems in urban areas'. *World Resource Review* 10:374–83.

——— 2004. 'Progress in heat-watch warning system technology'. *Bulletin of the American Meteorological Society* 85:1931–41.

Shrubsole, D. 2000. 'Flood management in Canada at the crossroads'. ICLR Research Paper Series, No. 5.

——— 2007. 'From structures to sustainability: A history of flood management strategies in Canada'. *International Journal of Emergency Management* 4 (2): 183–196.

Simonovic, S., and R. Carson. 2003. 'Flooding in the Red River Basin—Lessons from post flood activities'. *Natural Hazards* 28:345–65.

Smith, K. 1996. *Environmental hazards: Assessing risk and reducing disaster.* London, UK: Routledge.

Smith, S., G. Reuter, and M. Yau. 1998. 'The episodic occurrence of hail in central Alberta and the Highveld of South Africa'. *Atmosphere-Ocean* 36 (2): 169–78.

Smoyer, K., and D. Rainham. 2000. 'Heat-stress–related mortality in five cities in Southern Ontario: 1980–1996'. *International Journal of Biometeorology* 44:190–7.

Smoyer-Tomic, K., R. Kuhn, and A. Hudson. 2003. 'Heat wave hazards: An overview of heat wave impacts in Canada'. *Natural Hazards* 28:463–85.

Stenseth, N.C., O. Geir, J. Hurrell, and A. Belgrano, eds. 2004. *Marine ecosystems and climate variation.* Don Mills, ON: Oxford University Press.

Stephenson, D.B. 2000. 'Use of the "odds ratio" for diagnosing forecasting skill'. *Weather and Forecasting* 15:221–32.

Stewart, R.E., D. Bachand, R.R. Dunkley, A.C. Giles, B. Lawson, L. Legal, S.T. Miller, B.P. Murphy, M.N. Parker, B. Paruk, and M.K. Yau. 1995. 'Winter storms over Canada'. *Atmospheric-Ocean* 33:223–47.

Stewart, T.R., and C.M. Lusk. 1994. 'Seven components of judgmental forecasting skill: Implications for research and the improvement of forecasts. *Journal of Forecasting* 13:575–99. Reprinted in T. Connolly, H.R. Arkes, and K.R. Hammond, eds, *Judgment and decision making: An interdisciplinary reader*, 2nd ed. (New York: Cambridge University Press, 2000).

Stone, D., A. Weaver, and F. Zwiers. 2000. 'Trends in Canadian precipitation intensity'. *Atmosphere-Ocean* 38 (2): 321–47.

Stott, P.A., D.A. Stone, and M.R. Allen. 2004. 'Human contribution to the European heatwave of 2003'. *Nature* 432:610–13.

Street, R., D. Etkins, and D. Phillips. 1997. 'Weather impacts in Canada'. Workshop on the Social and Economic Impacts of Weather, Boulder, CO.

Strommel, H.G. 1966. 'The Great Blizzard of '66 on the northern Great Plains'. *Weatherwise* 19:189–207.

Stuart, R.A., and G.A. Isaac. 1999. 'Freezing Precipitation in Canada'. *Atmosphere-Ocean* 37 (1): 87–102.

Sutherland, A. 2006. 'Flash freeze puts city on ice'. *Montreal Gazette*, 19 January.

Szymczak, H., and T. Krishnamurti. 2006. 'Skill of synthetic superensemble hurricane forecasts for the Canadian maritime provinces'. *Meteorology and Atmospheric Physics* 93:147–63.

Taylor, N., D. Sills, J. Hanesiak, J. Milbrandt, P. McCarthy, C. Smith, and G. Strong. 2007. 'The Understanding Severe Thunderstorms and Alberta Boundary Layers Experiment (UNSTABLE)'. A report following the first science workshop, 18–19 April 2007, Edmonton, AL. *CMSO Bulletin SCMO* 35 (special issue): 20–8.

Thomalla, F., T. Downing, E. Spanger-Siegfried, G. Han, and J. Rockstrom. 2006. 'Reducing hazard vulnerability: Towards a common approach between disaster risk reduction and climate adaptation'. *Disasters* 30 (10): 39–48.

Thomas, M.K. 1971. 'A brief history of meteorological services in Canada, Part 1: 1839–1930'. *Atmosphere* 9 (1).

Thornes, J.E., and D.B. Stephenson. 2001. 'How to judge the quality and value of weather forecast products'. *Meteorological Applications* 8:307–14.

Tobin, G., and B. Montz. 1997. *Natural hazards*. New York: Guilford Press.

Tol, R.S. 2002. 'Estimates of the damage costs of climate change, Part 1: Benchmark Estimates'. *Environmental and Resource Economics* 21:47–73.

Tompkins, H. 2002. 'Climate change and extreme weather events: Is there a connection?' *Cicerone* 3:1–5.

Toth, Z., J. Desmarais, G. Brunet, P. Houtekamer, Y. Zhu, R. Wobus, R. Hogue, R. Verret, L. Wilson, B. Cui, G. Pellerin, B. Gordon, E. O'Lenic, and D. Unger. 2005. 'The North American Ensemble Forecast System (NAEFS)'. NAEFS Multinational Meeting Report.

Transport Canada. 2001. 'Trends in motor vehicle traffic collision statistics, 1988–1997'. Prepared by the Road Safety and Motor Vehicle Regulation Directorate, February.

Trenberth, K. 2005. 'Uncertainty in hurricanes and global warming'. *Science* 308:1753–4.

United Nations. 2004. *Living with risk: A global review of disaster reduction initiatives*. International Strategy for Disaster Reduction. www.e11th-hour.org/public/natural/living.with.risk.html.

USDOC (US Department of Commerce). 1985. 'National Hurricane Operations Plan'. NOAA, FCM-P12-1985, Washington DC.

Vancouver Sun. 2006. 'Tofino stays afloat after agreeing to truck in water'. www.canada.com/vancouversun/news/story.html?id=1a1a9ac8-1c71-4734-a57f-31d4248a34f7&k=44060.

Verret, R., F. Pithiios, L. Lefaivre, G. Pellerin, M.Klasa, P. Houtekamer, and L. Wilson. 2006. 'Spread-skill relationship in the Canadian ensemble prediction system'. Canadian Meteorological Centre report.

Vimont, D.J., and J.P. Kossin. 2007. 'The Atlantic Meridional Mode and hurricane activity'. *Geophyical Research Letters* 34:L07709, doi:10.1029/2007GL029683.

Vincent, L., and E. Mekis. 2006. 'Changes in daily and extreme temperature and precipitation indices for Canada over the twentieth century'. *Atmosphere-Ocean* 44 (2): 177–93.

Walsh, J.E., I. Shapiro, and T.L. Shy. 2005. 'On the variability and predictability of daily temperatures in the Arctic'. *Atmosphere-Ocean* 43 (3): 213–30.

Walsh, K. 2004. 'Tropical cyclones and climate change: Unresolved issues'. *Climate Research* 27:78–83.

Weather Doctor Almanac, The. 1999. 'Two flags flying'. www.islandnet.com/~see/weather/almanac/arc_1999/alm99oct.htm.

Weather Doctor, The. 2005. 'Newfoundland's Wreckhouse Winds'. www.islandnet.com/~see/weather/almanac/arc2005/alm05dec.htm.

Westeveld, J.S. 1996. 'Severe weather phobia: An exploratory study'. *Journal of Clinical Psychology* 52:509–15.

Westeveld, J.S., A. Less, T. Ansley, and H. Sook Yi. 2006. 'Severe-weather phobia'. *Bulletin of the American Meteorological Society* 87 (6): 747–9.

Wheaton, E., V. Wittrock, S. Kulshreshtha, G. Koshida, C. Grant, A. Chipanshi, and B. Bonsal. 2005. *Lessons learned from the Canadian drought years of 2001 and 2002: Synthesis report*. SRC Publication No. 11602–46E03. Ottawa: Agriculture and Agri-Food Canada.

Whiffen, B., P. Delannoy, and S. Siok. 2003. 'Fog: Impact on road transportation and mitigation options'. Presented at the Tenth World Congress and Exhibition on Intelligent Transportation Systems and Services, 16–20 November 2003, Madrid, Spain.

White, K., C. McKie, and J. Laverdiere. 1995. 'Human response to warning: The August 4th, 1994 Aylmer, Quebec tornado'. Report for the City of Aylmer, EPC, EC, and Regional Municipality of Ottawa-Carleton.

Wikipedia. 2007a. 'Alberta clipper'. http://en.wikipedia.org/wiki/Alberta_clipper.

——— 2007b. 'Chinook Wind'. http://en.wikipedia.org/wiki/Chinook_wind.

——— 2007c. 'Newfoundland hurricane of 1775'. http://en.wikipedia.org/wiki/Newfoundland_Hurricane_of_1775.

——— 2007d. 'Squamish Wind'. http://en.wikipedia.org/wiki/Squamish_(wind).

Wilson, L.J. 2001. 'Ensemble forecasting—The Canadian experience'. Paper presented at the WMO Commission for Basic Systems Workshop on the Use of Ensemble Prediction, Beijing, 16–20 October 2000. WMO Technical Document No. 1065.

Winton, M. 2006a. 'Amplified Arctic climate change: What does surface albedo have to do with it?' *Geophysical Research Letters* 33: L03701, doi: 10.1029/2005GL025244.

——— 2006b. 'Does the Arctic sea ice have a tipping point?' *Geophysical Research Letters* 33 (23): L23504, doi:10.1029/2006GL028017.

WMO (World Meteorological Organization). 1995. 'Meeting of experts to review the present status of hail suppression'. Report No. 26, South Africa, WMO/TD No. 764.

——— 2004. 'Weather forecasting technique considered as a sequence of standard processes from the forecaster's point of view'. Paper presented at a meeting of the Implementation Coordination Team on Data-Processing and Forecasting System, Geneva, Switzerland.

Zhang, X., L. Vincent, W. Hogg, and A. Niitsoo. 2000. 'Temperature and precipitation trends in Canada during the 20th century'. *Atmosphere-Ocean* 38 (3): 395–429.

Zhang, X., W.D. Hogg, and E. Mekis. 2001. 'Spatial and temporal characteristics of heavy precipitation events over Canada'. *Journal of Climate* 14 (9): 1923–36.

Index

adaptation: anticipatory, 176; autonomous, 176; climate change and, 181; mitigation and, 176–8; reactive, 176
advisories, 8, 14, 35–7
Agriculture and Agri-Food Canada, 109, 130
Akinremi, O., et al., 132
Alberta: summer weather in, 89, 90–2, 100; winter weather in, 51–3; *see also* Prairies
Alberta clippers, 49–50, 58
Alberta Low, 46, 78
Alberta Weather Centre, 39
alerts, 69, 79
Alexander, D., 170–1
'all-hazards approach', 180
ALIAWATCH, 39
American Meteorological Society, 177
Annan, Kofi, 3
Anthes, R.A., et al., 167
Arctic: climate change and, 102, 104; winter storm tracks in, 47; winter weather in, 55, 60
Arctic Bay, Nunavut, 106
Ashmore, P., and M. Church, 133
Atlantic Basin hurricane database (HURDAT), 143
Atlantic Canada: drought in, 128; floods in, 111, 113, 115; fog in, 122; smog in, 88; tropical cyclones and, 138–41, 144–5, 152–66
Atlantic Hurricane Seasonal Outlook, 151–2
Atlantic Meridional Model (AMM), 166–7
Atlantic Multidecal Oscillation (AMO), 167

Bailey, W.G., et al., 54
Bankoff, G., 171
Barrie tornado, 10, 96
black ice, 72

blizzards, 58–62; notable, 59–62; warning for, 37
'Blizzard of '88', 61
Bonsal, B., et al., 102, 134
Bonsal, B., and R. Lawford, 129
Bowyer, Peter, 171–2
British Columbia: drought in, 134; floods in, 111, 116; precipitation in, 105–6, 107, 120; smog in, 87–8; winter weather in, 10, 42, 53
Brooks, G.R., et al., 110
Brun, S.E., et al., 1
buoys: hurricanes and, 150–1

Calgary: hailstorms in, 10, 11
Canada-wide Public Alerting System (CPAS), 37
Canadian Centre for Emergency Preparedness, 179–80
Canadian Climate Impacts Scenarios (CCIS) Project, 74–5
Canadian Health Assessment, 103
Canadian Hurricane Centre, 139, 147–52; Information Statements of, 147; Response Zone of, 147–8
Canadian Institute for Climate Studies, 6–7
Canadian Lightning Detection Network (CLDN), 33–4, 100
Canadian Meteorological Centre (CMC), 25
Canadian Natural Hazards Assessment Project, 4–5, 9
Canadian Risk and Hazards Network (CRHNet), 14–15, 179–80
Canadian Wildland Fire Information System, 15
CanAlert, 35, 37
CANWARN, 38–9
Carter, A.O., et al., 96

Chicago: heat wave in, 80
China: flood in, 109–10
chinook, 51–3; arch, 53
'citizen forecasts', 38–9
climate change: drought and, 133–5; fog and, 133; natural climatic oscillations and, 15–19; precipitation and, 131–2; preparedness and, 179–81; severe weather and, 1–2, 8; summer weather and, 101–4; tropical cyclones and, 166–8; winter weather and, 73–5
Climate Severity Index (CSI), 6–9
clippers, 49–50, 58
cloud seeding, 91–2
Coastal Lows, 46, 78
cold: wind chill and, 67–72
cold wave advisory, 37
Colorado Climate Center, 131
Colorado Low, 46, 78
comma head, 43
costs: disasters and, 9–12, 179; drought and, 12, 129–30; hail and, 89–90
Crichton, D., 173
cyclones, 43–5; extratropical, 43–5; mid-latitude, 140; subtropical, 140; *see also* tropical cyclones

Danard, M., et al., 146–7
data: acquisition of, 25; analysis of, 27–8; assimilation of, 27–8
deaths: floods and, 110; fog and, 125, 126; heat and, 79, 102; lightning and, 101; smog and, 85; weather-related, 10–11
'Dirty Thirties', 130
Disaster Financial Assistance Arrangements (DFAA), 15, 179
disasters: costs of, 9–12, 179; hazards and, 3–5; number of, 11
Dolney, T., and S. Sheridan, 103
Doppler radar, 29–31
Dore, M.H., 179
downbursts, 96
dropsonde, 149, 150
drought, 105, 127–31; agricultural, 127; categories of, 127; climate change and, 133–5; costs of, 12, 129–30; definition of, 127; factors in, 128–9; hydrological, 127; indices of, 130–1; meteorological, 127; severe, 129–30; socio-economic, 127
'Drought Watch', 130

Dust Bowl, 128, 129
dust storm advisory, 36

Edmonton: tornado in, 10, 39, 99–100
El Niño–Southern Oscillations (ENSOs), 16, 93, 129, 151–2, 167
electromagnetic spectrum, 32
Emanuel, K., 167
emergency management, 179–80
Emergency Measures Organization (EMO), 35
Emergency Preparedness Act, 179
Empress of Ireland, 125–6
ensemble prediction system (EPS), 28
Environment Canada, 179–80; floods and, 119, 120; forecasting and, 25, 31, 38–9; vulnerability factors and, 4–5; warnings and advisories and, 8, 15, 42; wind chill and, 68–9
Escuminac Hurricane, 161–2
Etkin, D., et al., 13, 93, 94
Etkin, D., and M. Myers, 175
Europe: heat wave in, 80, 83, 102
extratropical transition (ET), 140
extreme cold weather alerts, 69
extreme heat alert, 79
eye, hurricane, 140, 145, 150

flash density, 101
floods, 109–20; causes and examples of, 113–16; flash, 113–14; management of, 116, 119–20; outburst, 115–16; significant, 117–19
Flood Damage Reduction Program (FDRP), 119, 120
fog, 120–6; aviation, 124–5; climate change and, 133; coastal, 125; definition of, 120; dispersal of, 126; frontal, 123–4; impacts of, 125–6; killer, 126; processes and types of, 123–4; radiation, 123–4, 125; upslope, 124; valley, 124
forecasting, 21–40; extended-range, 151–2; history of, 23–4; long-range, 34; models for, 26–8; procedure for, 24–8; quality of, 22–3; seasonal, 34, 151–2; tropical cyclones and, 147–52; volunteers and, 38–9
freezing drizzle, 65
freezing rain, 62–7; warning for, 36
frost, 71; warning for, 36
Fujita Scale, 95

Geis, D., 172
generalized environmental multiscale (GEM) model, 28
Geological Survey of Canada, 179–80
Geostationary Operational Environmental Satellites (GOES), 32
Global Atmospherics Inc., 33
Global Environmental Multiscale (GEM) Model, 147–8
global spectral model, 28
government: emergency management and, 179–81; mitigation and, 14–15
Great Blizzards, 62
'Great White Hurricane', 61
green roofs, 83
Groisman, P., et al., 56
Gulf of Alaska Low, 46
Gulf Stream, 138, 140
gustnadoes, 96

Hage Report, 39
hail, 77, 89–92; worst, 97–9
Hail Suppression Project, 91–2
Hatteras Low, 46, 78
hazards, 9–13; context of, 3–5
hazard index, 7
health: climate change and, 103; smog and, 85, 87
Hearn Morrow, B., 178
heat, 79–83; climate change and, 102–4
heat islands, 82–3
heat stress, 80, 83
heat waves, 79–82; significant, 81–2
heavy rain warning, 36
heavy snowfall warning, 37
Hudson Low, 78
humidity, 83–5; relative, 85
hurricane, 116; major, 142–3; as term, 141; vulnerability and, 171–2; *see also* tropical cyclones
Hurricane Able, 161
Hurricane Belle, 157
Hurricane Bertha, 157
Hurricane Beth, 145, 156
Hurricane Blanche, 156
Hurricane Bob, 157
Hurricane Celia, 156
Hurricane Daisy, 156
Hurricane Doria, 156
Hurricane Edna, 155

Hurricane Ella, 157
Hurricane Frederic, 157
Hurricane Epsilon, 144
Hurricane Gabrielle, 141, 144, 157, 158
Hurricane Ginny, 156, 162
Hurricane Gladys, 156
Hurricane Gustav, 30, 158
Hurricane Hazel, 117, 138, 144, 155, 161
Hurricane Helene, 155
Hurricane Hortense, 157, 162–3
Hurricane Juan, ix, 10, 146, 158, 163–6, 171–2
Hurricane Luis, 157
Hurricane Michael, 150, 157, 163
Hyogo Framework, 172

ice, black, 72
ice jams, 114–15
ice storms, 62–7; (1998), 10, 11, 12, 179, 63–4, 65–7
Independence Hurricane, 155, 158–9
Info-Smog, 87
Institute for Catastrophic Loss Reduction, 178, 179–80
insurance: disasters and, 179; drought and, 130; hail and, 89, 92
Insurance Bureau of Canada, 12
Intergovernmental Panel on Climate Change (IPCC), 18–19, 73, 102, 103, 131–2, 177
International Workshop on Tropical Cyclones, 167–8
Internet: warnings and, 37

Jaeger, C., et al., 173

Karl, T.R., and D.R. Easterling, 132
King, D., 171
Kocin, P.J., 61

Lambert, S.J., 73–4
lapse rate, 89
Lawson, B., 58, 60, 74
lightning, 33–4, 100–1; definition of, 100
Lower Fraser Valley Air Quality Monitoring Network, 87–8

McBean, G., 104
MacDougall, Lauchlan, 49
McIntyre, C., 5
Mackenzie Low, 46, 78

Manitoba: floods in, 10, 111, 114; smog in, 88; tornadoes in, 92; warnings in, 35; winter weather in, 42, 69; *see also* Prairies
maps: forecasting and, 27
marine wind warning, 36
Marion Bridge Doppler radar station, 30
Maybank, J., et al., 134
media: forecasting and, 22, 24
Meehl, G.A., and C. Tebaldi, 103
Mekis, E., and W. Hogg, 132
Meszaros, Jacqueline, 175
Meteorological Service of Canada (MSC), 15, 150–1; forecasting and, 23, 25, 28, 34
mitigation, 13–15; adaptation and, 176–8; definition of, 177; strategies for, 169, 178, 180–1; climate change and, 16, 103–4; numerical, 24, 26–8; thunderstorms and, 89; tropical storms and, 147–8, 151
multidecadal signal, 151–2
Muraca, G., et al., 123

names: hurricanes and, 143–4
National Adaptation Programs of Actions (NAPAs), 177
National Disaster Mitigation Strategy, 180
National Hurricane Center, 139, 143
National Science Foundation, 174–5
National Weather Service (NWS), 28
Natural Disaster Mitigation Strategy, 119
natural hazard risk, 3
Natural Resources Canada, 133, 134
Newark, M., 93
New Brunswick: floods in, 111, 113; tropical cyclones in, 146, 159; *see also* Atlantic Canada
Newfoundland: fog in, 122; tropical cyclones and, 138, 141, 144–5, 158–9; winter weather in, 48–9; *see also* Atlantic Canada
Newton, J., 177
Niagara Falls: freezing of, 71–2
Niagara snowstorm, 42
Nkemdirim, L., and L. Weber, 129
Noel (post-tropical storm), 144–5, 146, 171
nor-easters, 50–1
North: climate change and, 102, 104; forecasting and, 39–40; hazards and, 180; winter storm tracks in, 47; winter weather in, 55, 60
North American Drought Monitor (NADM), 130

North Atlantic Oscillation, 16, 17, 167
Nova Scotia: tropical cyclones and, 144–5, 146, 159, 163–6, 171–2; winter weather in, 42, 55, 56; *see also* Atlantic Canada
numerical weather prediction (NWP), 27
Nunavut: forecasting in, 39–40; winter storm tracks in, 47: *see also* Arctic; North

Ontario: floods, in, 111; summer weather in, 79, 80–1, 83, 85; tornadoes in, 93; winter weather in, 42, 69
Ontario Medical Association, 85
oscillations, climatic, 15–19, 167
outbursts, 115–16
outdoor mobility index, 8
ozone, 85–8

Pacific Low, 78
Palmer Drought Severity Index (PDSI), 130–1
Paul, 94–5
Pelmorex Forecast Engine (PFE), 22
Perfect Storm, 162
Pielke, R.A., et al., 167
Pine Lake tornado, 10
'Pineapple Express', 106
pollution: fog and, 126
Prairies: air quality in, 88; climate change and, 132; drought in, 128–30; hail and, 90–1; precipitation in, 105, 107, 108–9; tornadoes in, 93–100; *see also* specific provinces
precipitation, 105–35; climate change and, 18, 131–4; convectional, 108–9; flooding and, 113–14; frontal, 109; heavy, 106–9; orographic, 109; processes of, 108–9; tropical cyclones and, 139–40, 145; *see also* snow
preparedness, 13–14, 170, 180
Prince Rupert, BC, 105–6
psychological index, 7
Public Safety Canada (PSC), 14, 179–80

Quebec: floods in, 111, 113, 114; smog in, 87; winter weather in, 56

radar: forecasting and, 28–31, 149–50
Raddatz, R.L., and J.M. Hanesiak, 109
Radio Amateur Educational Society (RAES), 39

radiosondes, 25
rain: *see* precipitation
recovery, 13–14, 172
Red River flood, 111, 114, 119–20, 179
Red River Floodway, 15, 119–20, 178
Redner, S., and M. Petersen, 103
Regina: humidity in, 84; tornado in, 96, 99
Renn, O., 175
resilience, 172
response, 13–14
Ripley, E., 132
risk: cognitive factors in, 174; perception of, 173–6; reduction of, 177–8; situational factors in, 174; vulnerability and, 170
'risk burdens', 3
'risk triangle', 173
roads: winter and, 42, 72
Roberts, E., et al., 129

Saffir-Simpson Hurricane Scale, 141
Saguenay flood, 10, 114, 179
satellite imagery, 31–2; cold and, 72; tropical cyclones and, 149–50
Saxby Gale, 146, 155, 159
sea surface temperatures (SSTs), 17, 138
Seguin, J., 103
seiches, 116
severe thunderstorm watch/warning, 36
severe weather: classification of, 5–9; climate change and, 1–2; definition of, 5–9; number of events, 2
Shrubsole, D., 116, 119
smog, 85–8; climate change and, 103
Smoyer, K., and D. Rainham, 80
Smoyer-Tomic, K., et al., 80
snow, 53–7; blowing, 58
snow cover, 41
snowmelt runoff, 113
snow squalls, 45
snowstorms, 42; notable, 56–7; *see also* blizzards
squall lines, 45
squamish, 53
Standard Precipitation Index (SPI), 130, 131
Stewart, T.R., and C.M. Lusk, 23
Stone, D., et al., 132, 133
storms: coastal, 116; extratropical, 137–40; lake-effect, 45; multicell, 88; post-tropical, 139, 144–5, 146, 171; supercell, 88–9; tropical, 141; winter, 42–7; *see also* blizzards; ice storms; snowstorms; thunderstorms; tropical cyclones
storm surges, 116, 141–2, 146–7
storm tracks: summer, 78–9; winter, 45–7
stormwater runoff, 116
Street, R., et al., 3
summer discomfort index, 7
summer weather, 77–104; climate change and, 101–4
synthetic aperture radar (SAR), 149–50
system simulation experiment (SSE), 28

thunderstorms, 33–4, 88–9; air mass, 88; tornadoes and, 95–6
Tobin, G., and B. Montz, 173–4
Tofino, BC, 127
Tompkins, H., 167
'tornado alleys', 93
tornadoes, 77, 92–100; categorizing, 95; climate change and, 104; tropical cyclones and, 147; watch/warning for, 36; worst, 97–9
Toronto: cold in, 69; heat in, 80–1, 83
tropical cyclones, 137–68; classification of, 138, 141–3; climate change and, 166–8; definition of, 139; forecasting of, 147–52; impacts of, 144–7; naming of, 143–4; notable Canadian, 152–66; processes of, 139–41; tracks of, 152–3
tropical depression, 141
tropical storm, 141
Typhoon Freda, 138, 156

Understanding Severe Thunderstorms and Alberta Boundary Layers Experiment (UNSTABLE), 89
United Nations Development Program (UNDP), 176–7
urban heat island, 82–3
urban runoff, 116

Vimont, D.J., and J.P. Kossin, 166–7
Vincent, L., and E. Mekis, 102
'vivid messages', 175
vulnerability, 3–5, 170–2; climate change and, 181; delinquent, 171; factors in, 4–5; newly generated, 171; residential,

171; risk and, 170, 173–6; socio-economic, 170; technological, 170; types of, 170–1

warnings, 8, 15, 23, 35–7, 42; blizzard, 37; heat, 79; risk and, 174; smog, 87; thunderstorm, 88; tornado, 88, 92, 96; wind, 36, 37
weather advisories, 8, 35–7
Weather Network, 22, 37
Weatheradio, 39–40
weather watches, 8, 15, 35–7
White, K., et al., 96
'White Juan', 56
whiteout, 58
White River, Ontario, 69
wind: chinook, 51–3; sustained, 140; tornadoes and, 95; tropical cyclones and, 139–40, 141–2, 144–5; winter, 45

wind chill, 67–72; index, 69; warning for, 37
Winnipeg: cold in, 69; 1993 flood in, 10
winter discomfort index, 7
winter storm warning, 37
winter weather, 41–79; causes of, 43–5; climate change and, 73–5
World Conference on Disaster Reduction (WCDR), 172
World Meteorological Centre, 139, 143
World Meteorological Organization (WMO), 25
Wreckhouse winds, 45, 48–9

Zhang, X., et al., 73, 102, 132